Optimal Control Applications in Electric Power Systems

MATHEMATICAL CONCEPTS AND METHODS IN SCIENCE AND ENGINEERING

Series Editor: Angelo Miele
Mechanical Engineering and Mathematical Sciences
Rice University

Recent volumes in this series:

22 **APPLICATIONS OF FUNCTIONAL ANALYSIS IN ENGINEERING** • *J. L. Nowinski*

23 **APPLIED PROBABILITY** • *Frank A. Haight*

24 **THE CALCULUS OF VARIATIONS AND OPTIMAL CONTROL: An Introduction** • *George Leitmann*

25 **CONTROL, IDENTIFICATION, AND INPUT OPTIMIZATION** • *Robert Kalaba and Karl Spingarn*

26 **PROBLEMS AND METHODS OF OPTIMAL STRUCTURAL DESIGN** • *N. V. Banichuk*

27 **REAL AND FUNCTIONAL ANALYSIS, Second Edition Part A: Real Analysis** • *A. Mukherjea and K. Pothoven*

28 **REAL AND FUNCTIONAL ANALYSIS, Second Edition Part B: Functional Analysis** • *A. Mukherjea and K. Pothoven*

29 **AN INTRODUCTION TO PROBABILITY THEORY WITH STATISTICAL APPLICATIONS** • *Michael A. Golberg*

30 **MULTIPLE-CRITERIA DECISION MAKING: Concepts, Techniques, and Extensions** • *Po-Lung Yu*

31 **NUMERICAL DERIVATIVES AND NONLINEAR ANALYSIS** • *Harriet Kagiwada, Robert Kalaba, Nima Rasakhoo, and Karl Spingarn*

32 **PRINCIPLES OF ENGINEERING MECHANICS Volume 1: Kinematics—The Geometry of Motion** • *Millard F. Beatty, Jr.*

33 **PRINCIPLES OF ENGINEERING MECHANICS Volume 2: Dynamics—The Analysis of Motion** • *Millard F. Beatty, Jr.*

34 **STRUCTURAL OPTIMIZATION Volume 1: Optimality Criteria** • *Edited by M. Save and W. Prager*

35 **OPTIMAL CONTROL APPLICATIONS IN ELECTRIC POWER SYSTEMS** *G. S. Christensen, M. E. El-Hawary, and S. A. Soliman*

A Continuation Order Plan in available for this series. A continuation order will bring delivery of each new volume immediately upon publication. Volumes are billed only upon actual shipment. For further information please contact the publisher.

Optimal Control Applications in Electric Power Systems

G. S. Christensen

University of Alberta
Edmonton, Alberta, Canada

M. E. El-Hawary

Technical University of Nova Scotia
Halifax, Nova Scotia, Canada

and

S. A. Soliman

Ain Shams University
Cairo, Egypt

PLENUM PRESS • NEW YORK AND LONDON

Library of Congress Cataloging in Publication Data

Christensen, G. S. (Gustav S.)
Optimal control applications in electric power systems.

(Mathematical concepts and methods in science and engineering; 35)
Includes bibliographies and index.
1. Electric power systems—Control. I. El-Hawary, M. E. II. Soliman, S. A. III. Title. IV. Series.
TK1005.C47 1987 621.319 87-7712
ISBN 0-306-42517-3

A Division of Plenum Publishing Corporation
233 Spring Street, New York, N.Y. 10013

Printed in the United States of America

Preface

Significant advances in the field of optimal control have been made over the past few decades. These advances have been well documented in numerous fine publications, and have motivated a number of innovations in electric power system engineering, but they have not yet been collected in book form. Our purpose in writing this book is to provide a description of some of the applications of optimal control techniques to practical power system problems.

The book is designed for advanced undergraduate courses in electric power systems, as well as graduate courses in electrical engineering, applied mathematics, and industrial engineering. It is also intended as a self-study aid for practicing personnel involved in the planning and operation of electric power systems for utilities, manufacturers, and consulting and government regulatory agencies.

The book consists of seven chapters. It begins with an introductory chapter that briefly reviews the history of optimal control and its power system applications and also provides an outline of the text. The second chapter is entitled "Some Optimal Control Techniques"; its intent is to introduce fundamental concepts of optimal control theory that are relevant to the applications treated in the following chapters. Emphasis is given to clear, methodical development rather than rigorous formal proofs. Topics discussed include variational calculus, Pontryagin's maximum principle, and geometric methods employing functional analysis. A number of solved examples are included to illustrate the techniques.

"Dynamic Optimal Load Flow" is the title of Chapter 3. This chapter deals with the problem of economic scheduling of electric power systems including thermal and hydrothermal generation resources. The network model utilized is the exact load flow formulation. The case of hydroplants with essentially constant head is addressed. Computational approaches are detailed and sample test cases are provided.

"Economic Coordination of Hydrothermal-Nuclear Systems" is treated in Chapter 4. In this chapter the details of optimal economic operation strategies for systems including nuclear capability are discussed. For the sake of clarity in presentation, the electric network is represented by an active power balance model.

The problem of optimally controlling the tie-line flow in an interconnected system so as to maintain acceptable frequency variations in response to load variations is detailed in Chapter 5, "Optimal Tie-Line Control." Computational examples are included in this chapter.

"Optimal Control of Turboalternators" is the title of Chapter 6. Open loop and feedback optimal control strategies for turboalternator stabilization are discussed in Chapter 6. Torque and voltage variables are taken as the control inputs in both cases. Computational examples are provided to compare the performance of the different methods of approach.

Chapter 7 summarizes the developments discussed in this monograph. Directions for further improvements in model complexity versus worth are highlighted.

G. S. Christensen
Edmonton, Alberta, Canada
M. E. El-Hawary
Halifax, Nova Scotia, Canada
S. A. Soliman
Cairo, Egypt

Acknowledgments

During the planning and writing of this book, we have incurred indebtedness to many people. We wish to acknowledge continuing encouragement received from Dr. C. R. James, Chairman of the Department of Electrical Engineering at the University of Alberta. We are grateful to our many friends and colleagues and, in particular, to Dr. D. H. Kelly of the University of Alberta, Dr. M. Abdelhalim of the University of Alexandria, Egypt, Mr. Y. Chen, Chinese Scholar, Dr. R. S. Rao, and Dr. A. Shamaly. We wish to express our thanks to Ms. Barbara J. Peck, Director, Canterbury Executive Services, for a neatly done job of typing the many drafts of the manuscript. This work would not have been possible without the patience and understanding of our wives, Dr. Penelope Christensen, Dr. Ferial El-Hawary, and Laila Soliman.

G.S.C.
M.E.E.
S.A.S.

Contents

1. Introduction

1.1. The Role of Optimization and Computational Mathematics ... 1
1.2. Outline of the Book ... 2

2. Some Optimal Control Techniques

2.1. Introduction ... 5
2.2. Calculus of Variations ... 5
2.2.1. Unconstrained Optimization ... 5
2.2.2. Constrained Optimization ... 9
2.3. Pontryagin's Maximum Principle ... 10
2.4. The Functional Analytic Optimization Technique ... 14
2.4.1. Norms ... 15
2.4.2. Inner Product (Dot Product) ... 15
2.4.3. Transformations ... 17
2.4.4. A Minimum Norm Theorem ... 18
References ... 20

3. Dynamic Optimal Load Flow

3.1. Introduction ... 21
3.2. The All-Thermal OPF ... 22
3.3. The Hydrothermal OPF ... 25
3.4. A Successive Approximation Solution ... 27
3.5. A Newton-Iterative Solution ... 33
3.5.1. The Jacobian ... 34
3.5.2. Initial Guess Estimation ... 37
3.6. Computational Performance ... 39
3.6.1. Successive Approximation Algorithm ... 39
3.6.2. Newton's Method ... 44

3.7. Conclusions ... 52
References ... 53

4. Economic Coordination of Hydrothermal–Nuclear Systems
4.1. Introduction ... 55
4.2. Problem Formulation ... 56
4.3. The Optimization Procedure ... 58
4.4. The Optimal Solution ... 62
4.5. A Feasible Multilevel Approach ... 65
4.6. Conclusions and Comments ... 66
Appendix 4.A ... 67
Appendix 4.B ... 67
References ... 68

5. Optimal Tie-Line Control
5.1. Introduction ... 69
5.2. Load Frequency Control of Interconnected Power Systems 70
5.2.1. Mathematical Model ... 71
5.2.2. Problem Formulation ... 72
5.2.3. Optimal Solution ... 74
5.2.4. Solution Technique ... 77
5.2.5. Practical Example ... 78
5.3. Optimum Load Frequency Control of Multiarea Interconnected Power Systems ... 82
5.3.1. Problem Formulation ... 84
5.3.2. Optimal Solution ... 86
5.3.3. A Sudden Loss or Gain of Load at One Area ... 88
5.3.4. A Loss of a Tie-Line between Two Areas ... 91
5.3.5. A Sudden Loss of Generation at One Area ... 94
5.4. Optimal Load Frequency Control with Governor Backlash 99
5.4.1. Mathematical Model ... 100
5.4.2. Problem Formulation ... 101
5.4.3. Solution Technique ... 103
5.4.4. Practical Example ... 106
5.5. Conclusions ... 108
References ... 108

6. Optimal Control of Turboalternators
6.1. Introduction ... 111

6.2. Optimal Control of a Large Turboalternator 111
6.2.1. Mathematical Model 112
6.2.2. A Minimum Norm Formulation 114
6.2.3. The Optimal Solution 116
6.2.4. Solution Technique 117
6.2.5. Computational Example 118
6.3. Optimal Control of Two Interconnected Turbogenerators ... 119
6.3.1. Differential Equations of the System and Their Simplifications 121
6.3.2. Mathematical Model 129
6.3.3. Mathematical Solution 133
6.3.4. The Optimal Solution 137
6.3.5. Computation Techniques 142
6.4. Realistic Feedback Control of Two Interconnected Turbogenerators 147
6.4.1. Mathematical Model 152
6.4.2. Mathematical Solution 159
6.4.3. The Optimal Solution 165
6.4.4. Computational Technique 171
6.5. Conclusions 182
Appendix 6.A 184
Appendix 6.B: Steady-State Values and Machine Constants 186
Appendix 6.C: Steady-State Values and Machine Constants 187
Appendix 6.D: Steady-State Values and Machine Constants 188
References 190

7. Conclusion
7.1. Summary 191
7.2. Future Work 191

Index 193

1

Introduction

1.1. The Role of Optimization and Computational Mathematics

During the past two and a half decades the theory of optimal control has found an increasing number of applications in electric power systems engineering. This is in large measure due to the advent of large-scale digital computers. Problems that earlier were quite impossible to solve numerically owing to time constraints can now be solved with relative ease. However, as is pointed out in, for example, Chapter 3 of this book, there are still problems in electric power systems engineering that are simply too large for economic solution even by large digital computers. Hence, there is a need for new and faster algorithms than the ones currently available.

The general field of power systems engineering is continually dependent in its development on a balanced blend of engineering operating experience and the skilled adoption of mathematical advances and discoveries. In this regard it is no different from other disciplines of engineering and applied science. The strong influence of newly developed optimization and computational techniques in the area can be partly judged by inspection of titles of published work over the past few decades.

Bellman's development of the principle of optimality and dynamic programming in the 1950s and the publication of his two early classic books gave quite an impetus to problems involving discrete and discontinuous variables.

The development of the Kuhn–Tucker theorem in 1951 and its introduction to the optimization literature contributed to the advances in formulating problems including inequality constraints. Indeed, the varied powerful nonlinear programming procedures continue to influence developments in the area of electric power systems engineering.

The introduction of Pontryagin's maximum principle in the late 1950s and early 1960s resulted in the solution of several problems within electric power systems by this method. As is the case with dynamic, linear, and nonlinear programming, the maximum principle continues to be one of the essential tools of optimization used in our types of problems.

The adaptation of functional analysis concepts to optimization leads to the emergence of the abstract minimum norm problems. These developments can be traced back to the late 1950s. Application of these powerful techniques to many optimal operation problems dates to the early 1970s.

1.2. Outline of the Book

The body of this book consists of Chapters 2–6. Chapter 2 provides background preparation and provides us with the tools necessary to define optimization problems. The remaining chapters provide presentation of problems and discussion. We shall now review the following chapters.

Chapter 2 considers the mathematical basis for studies and results to follow. The first important topic reviewed is unconstrained optimization by aid of the calculus of variations. Two example problems are presented and solved. Next constrained optimization is considered and Lagrange multipliers are introduced. The following section reviews Pontryagin's maximum principle and two examples are presented and solved. The last section in Chapter 2 reviews functional analytic concepts and concludes by introducing a form of the powerful minimum norm formulation.

Chapter 3 presents dynamic optimal load flow. The first section of that chapter introduces the problem and explains the difference between the static and dynamic cases. The second section describes the all-thermal optimal power flow problem, while the third section presents hydrothermal optimal power flow in detail. A successive approximation solution to the last-mentioned problem is presented next. A Newton-iterative solution to this problem is then given in Section 3.5. These last two iterative methods are then applied in turn to several problems of varying complexity. The results obtained are compared and discussed.

Chapter 4 discusses the problem of economic coordination of hydro-thermal–nuclear systems. The organization of the material in this chapter is as follows: in Section 4.2 the problem is defined and the development of the system's model is presented; in Section 4.3 the optimization problem is reformulated in the context of functional analysis; Section 4.4 presents the solution to the problem; and Section 4.5 proposes an iterative method for its implementation.

Chapter 5 considers the problem of optimal tie-line control. The first section of Chapter 5 is concerned with the design of a linear regulator for the problem of load frequency control (LFC) of interconnected power systems. The control is specified to be of proportional-plus-integral (PI) form and is only a function of the measurable states. The load frequency control problem is formulated as a parameter optimization problem by employing the Pontryagin minimum principle. The second section of this chapter is concerned with the optimum load frequency control of multiarea interconnected power systems. The control law in that section is assumed to have a proportional and/or integral form and is a function only of the measurable states (present-day practice). The third section is concerned with the design of an optimal controller for a system whose dynamical equations contain a backlash element; this design is applied to the problem of load frequency control of a single area steam power system.

In Chapter 6 our interest is focused on the solution of the problem of optimal control of synchronous turboalternators. In the first section of this chapter the optimal torque and voltage control of a large turbogenerator connected to an infinite bus is found by using the minimum norm formulation. In the second section the optimal control of a system with two identical interconnected turbogenerators connected to an infinite bus is considered. Control of the generators is effected through control of field voltages and turbine torques. Finally, in the last section a realistic feedback control of two interconnected turbogenerators is considered.

There are continuing changes in power system operating conditions and these require reevaluation of the methods used. Opportunities for improvements continue to exist for efforts to assure security and provide adequate control in the system. Chapter 7 is devoted to a summary and brief outline of certain directions for future research needs.

2

Some Optimal Control Techniques

2.1. Introduction

The intent of this chapter is to introduce fundamental concepts of optimal control theory that are relevant to the applications treated in the following Chapters. Emphasis will be given to clear, methodical development rather than rigorous formal proofs. The fundamental conditions of the variational calculus for dynamic system optimization are discussed in Section 2.2. A brief summary of Pontryagin's maximum principle is given in Section 2.3. In Section 2.4 the functional analytic technique of formulating optimization problems in the minimum norm form is dealt with. A brief treatment of some basic concepts from functional analysis is given prior to stating one powerful version of minimum norm problems.

2.2. Calculus of Variations

2.2.1. Unconstrained Optimization

The calculus of variations is concerned with the determination of extrema (maxima and minima) or stationary values of functionals. A functional can be defined as a function of several other functions. The calculus of variations is a powerful method for the solution of problems in optimal economy operation of power systems. In this section we introduce the subject of variational calculus through a derivation of the Euler equations and associated transversality conditions.

A simple problem in the theory of calculus of variations with no constraints can be stated as follows: Find a function $x(t)$ that minimizes the following objective function J:

$$J = \int_{t_0}^{t_f} L[x(t), \dot{x}(t), t]\, dt \tag{2.1}$$

We assume that $x^*(t)$ is an optimal and construct a family of functions $x(t)$ that includes $x^*(t)$:

$$x(t) = x^*(t) + \varepsilon\eta(t), \qquad \dot{x}(t) = \dot{x}^*(t) + \varepsilon\dot{\eta}(t) \tag{2.2}$$

The parameter ε is a real positive parameter. With the above equation, the functional of equation (2.1) becomes

$$J(\varepsilon) = \int_{t_0}^{t_f} L[x^*(t) + \varepsilon\eta(t), \dot{x}^*(t) + \varepsilon\dot{\eta}(t), t]\, dt \tag{2.3}$$

The cost functional in equation (2.3) is minimum when $x(t) = x^*(t)$, or when $\varepsilon = 0$. In other words for the cost functional in equation (2.3) to be a minimum

$$\left.\frac{dJ(\varepsilon)}{d\varepsilon}\right|_{\varepsilon=0} = 0 \tag{2.4}$$

from the principles of calculus of variations we have

$$\frac{dJ(\varepsilon)}{d\varepsilon} = \frac{\partial J(\varepsilon)}{\partial x(t)} \cdot \frac{dx(t)}{d\varepsilon} + \frac{\partial J(\varepsilon)}{\partial \dot{x}(t)} \cdot \frac{d\dot{x}(t)}{d\varepsilon} + \frac{\partial J(\varepsilon)}{\partial t}\frac{dt}{d\varepsilon} \tag{2.5}$$

By performing the differentiation inside the integration sign, we obtain

$$0 = \int_{t_0}^{t_f} [L_x\eta(t) + L_{\dot{x}}\dot{\eta}(t) + L_t 0]\, dt \tag{2.6}$$

where

$$L_x = \frac{\partial L(x, \dot{x}, t)}{\partial x} \tag{2.7}$$

$$L_{\dot{x}} = \frac{\partial L(x, \dot{x}, t)}{\partial \dot{x}} \tag{2.8}$$

$$L_t = \frac{\partial L(x, \dot{x}, t)}{\partial t} \tag{2.9}$$

Integrating the second term in equation (2.6) by parts we obtain

$$0 = \int_{t_0}^{t_f} \left(L_x - \frac{d}{dt}L_{\dot{x}}\right)\eta(t)\, dt + [L_{\dot{x}}\eta(t)]_{t_0}^{t_f} \tag{2.10}$$

Equation (2.10) is equal to zero if and only if

$$L_x - \frac{d}{dt}L_{\dot{x}} = 0 \tag{2.11}$$

$$[L_{\dot{x}}\eta(t)]_{t_0}^{t_f} = 0 \tag{2.12}$$

Equation (2.11) is called the Euler equation of the variational problems and is a necessary condition for an extremum. Equation (2.12) is the transversality condition.

For a two-fixed-end-points problem we have

$$L_{\dot{x}(t_0)} \neq 0, \qquad L_{\dot{x}(t_f)} \neq 0 \tag{2.13a}$$

$$\eta(t_0) = \eta(t_f) = 0 \tag{2.13b}$$

because at these two points $x(t_0) = x^*(t_0)$ and $x(t_f) = x^*(t_f)$.

For a free-end-point problem, we have

$$L_{\dot{x}(t_f)} = 0, \qquad \eta(t_f) \neq 0 \tag{2.14}$$

Example 2.1. Find the extremal $x(t)$ in each of the following cases:

(a) $$J = \int_0^1 [\tfrac{1}{2}\dot{x}^2(t) + 3x(t)\dot{x}(t) + 2x^2(t) + 4x(t)]\, dt$$

with $x(0) = 1$; $x(1) = 4$.

(b) $$J = \int_0^1 [x^2(t) + \dot{x}^2(t)]\, dt$$

with $x(0) = 1$; $x(1)$ is free.

Solution:

(a) $$L(x, \dot{x}, t) = \tfrac{1}{2}\dot{x}^2(t) + 2x(t)\dot{x}(t) + 4x(t) + 2x^2(t)$$

$$L_x = \frac{\partial L(x, \dot{x}, t)}{\partial x} = 2\dot{x}(t) + 4x(t) + 4$$

$$L_{\dot{x}} = \frac{\partial L(x, \dot{x}, t)}{\partial \dot{x}} = \dot{x}(t) + 2x(t)$$

Substituting in equation (2.11) we obtain the following Euler equation:

$$2\dot{x}(t) + 4x(t) + 4 - \frac{d}{dt}[\dot{x}(t) + 2x(t)] = 0$$

or

$$\ddot{x}(t) - 4x(t) = 4$$

By solving the above second-order differential equation, we obtain

$$x(t) = C_1 \exp(2t) + C_2 \exp(-2t) + 1$$

To determine C_1 and C_2 we must apply the boundary conditions

$$x(t = 0) = 1, \qquad x(t = 1) = 4$$

$$1 = C_1 + C_2 + 1$$

or

$$C_1 + C_2 = 0 \tag{a}$$

$$4 = C_1 \exp(2) + C_2 \exp(-2) + 1$$

or

$$C_1 \exp(2) + C_2 \exp(-2) = 3 \tag{b}$$

We can now solve equations (a) and (b) for C_1 and C_2:

$$C_1 = 0.652, \qquad C_2 = 1.348$$

Therefore, the extremal curve, which satisfies the given boundary conditions, is

$$x(t) = 0.652 \exp(2t) + 1.348 \exp(-2t) + 1$$

(b)

$$L(x, \dot{x}, t) = x^2(t) + \dot{x}^2(t)$$

$$L_x = 2x(t)$$

$$L_{\dot{x}} = 2\dot{x}(t)$$

Substituting into equation (2.11) we obtain the following Euler equation:

$$2x(t) = -\frac{d}{dt}[2\dot{x}(t)] = 0$$

or

$$\ddot{x}(t) - x(t) = 0$$

The solution of the above equation is given by

$$x(t) = C_1 \exp(t) + C_2 \exp(-t)$$

To determine C_1 and C_2 we must apply the transversality and the boundary conditions. For free end point

$$L_{\dot{x}(t_f)}\big|_{t_f=1} = 0$$

or

$$2\dot{x}(t_f) = 0$$

$$0 = C_1 \exp(1) + C_2 \exp(-1) \tag{a}$$

Also, we have the following boundary condition:

$$x(t = 0) = 1$$

$$0 = C_1 + C_2 \tag{b}$$

We can now solve for C_1 and C_2 from the simultaneous equations (a) and (b) to obtain $C_1 = 0.1192$ and $C_2 = 0.88079$. Therefore the extremal curve that satisfies the given boundary conditions is

$$x(t) = 0.1192 \exp(t) + 0.88079 \exp(-t)$$

2.2.2. Constrained Optimization

The problem examined in Section 2.2.1 excluded consideration of optimal control problems having constraint relationships between the scalar elements of the state trajectory, which occurs in many physical problems. The problem including such a constraint can be formulated as follows: Find the function $x(t)$ that minimizes the following cost functional:

$$J = \int_{t_0}^{t_f} L[x(t), \dot{x}(t), t]\, dt \tag{2.15}$$

Subject to satisfying the following constraint:

$$\int_{t_0}^{t_f} g[x(t), \dot{x}(t), t]\, dt = 0 \tag{2.16}$$

An example of this problem that occurs in the power system area is the minimization of the fuel cost of a power plant subject to satisfying the active power balance equation for the system.

We can form an augmented cost functional by augmenting the cost functional of equation (2.15) by equation (2.16) via Lagrange's multiplier λ,

$$\tilde{J} = \int_{t_0}^{t_f} \tilde{L}[x(t), \dot{x}(t), t]\, dt \tag{2.17}$$

where

$$\tilde{L}(\cdot) = L(\cdot) + \lambda g(\cdot) \tag{2.18}$$

As a result we obtain the following modified Euler equation:

$$\tilde{L}_{x(t)} - \frac{d}{dt}(\tilde{L}_{\dot{x}(t)}) = 0 \tag{2.19}$$

or

$$(L_{x(t)} + g_{x(t)}) - \frac{d}{dt}(L_{\dot{x}(t)} + \lambda g_{\dot{x}(t)}) = 0 \tag{2.20}$$

Example 2.2. We wish to maximize

$$J = \int_{-1}^{1} x(t)\, dt$$

Subject to satisfying

$$\int_{-1}^{1} [1 + \dot{x}^2(t)]^{1/2}\, dt = 1$$

The augmented cost functional is given by

$$\tilde{J} = \int_{-1}^{1} \{x(t) + \lambda[1 + \dot{x}^2(t)]^{1/2}\}\, dt$$

$$\tilde{L}(x, \dot{x}, t) = x(t) + \lambda[1 + \dot{x}^2(t)]^{1/2}$$

$$\tilde{L}_{x(t)} = \frac{\partial \tilde{L}(\cdot)}{\partial x(t)} = 1$$

$$\tilde{L}_{\dot{x}(t)} = \frac{\partial \tilde{L}(\cdot)}{\partial \dot{x}(t)} = \frac{\lambda \dot{x}(t)}{[1 + \dot{x}^2(t)]^{1/2}}$$

Substituting into equation (2.19), one obtains the following Euler equation:

$$1 - \lambda \frac{d}{dt} \frac{\dot{x}(t)}{[1 + \dot{x}^2(t)]^{1/2}} = 0$$

or

$$\frac{\dot{x}(t)}{[1 + \dot{x}^2(t)]^{1/2}} = \frac{1}{\lambda} t + C$$

The solution of the above equation is given by

$$(x - x_1)^2 + (t - t_1)^2 = r^2$$

where the parameters x_1, t_1, and r are chosen to satisfy the boundary conditions.

2.3. Pontryagin's Maximum Principle

Let $\mathbf{u}(t)$ be an admissible control and $\mathbf{x}(t)$ be the corresponding trajectory of the system described by

$$\dot{\mathbf{x}}(t) = (\mathbf{x}(t), \mathbf{u}(t), t) \tag{2.21}$$

Let $\mathbf{x}(t_0) = \mathbf{x}(0)$; t_0 and t_f are specified and $\mathbf{x}(t_f)$ is free. The necessary conditions for $\mathbf{u}(t)$ to be an optimal control, i.e., to be the control that takes $\mathbf{x}(t)$ from $\mathbf{x}(0)$ to some state $\mathbf{x}(t_f)$ while minimizing the functional J

$$J = G[\mathbf{x}(t_f, t_f] + \int_{t_0}^{t_f} L[\mathbf{x}(t), \mathbf{u}(t), t]\, dt \tag{2.22}$$

are as follows:

(1) There exists a function or vector $\boldsymbol{\lambda}(t)$ such that $\mathbf{x}(t)$ and $\boldsymbol{\lambda}(t)$ are the solutions of the following equations:

$$\dot{\mathbf{x}}(t) = \frac{\partial H}{\partial \boldsymbol{\lambda}(t)} \tag{2.23}$$

$$\dot{\boldsymbol{\lambda}}(t) = -\frac{\partial H}{\partial \mathbf{x}(t)} \tag{2.24}$$

Subject to the boundary conditions given by

$$\mathbf{x}(t_0) = \mathbf{x}(0) \tag{2.25}$$

$$\boldsymbol{\lambda}(t_f) = \left.\frac{\partial G(\cdot)}{\partial \mathbf{x}(t)}\right|_{t=t_f} \quad \text{at } x(t) = x(t_f) \tag{2.26}$$

where the function H is a scalar function, which is called the Hamiltonian and is given by

$$H[\mathbf{x}(t), \mathbf{u}(t), \boldsymbol{\lambda}(t), t] = L[\mathbf{x}(t), \mathbf{u}(t), t] + \boldsymbol{\lambda}^T(t) f[\mathbf{x}(t), \mathbf{u}(t), t] \tag{2.27}$$

(2) The functional $H[\mathbf{x}(t), \mathbf{u}(t), \boldsymbol{\lambda}(t), t]$ has a local minimum at

$$\frac{\partial H}{\partial \mathbf{u}(t)} = 0 \tag{2.28}$$

In many practical problems, there are inequality constraints on the admissible control and states and these constraints must be taken into account. As a result (2.28) is no longer applicable. The basic contribution of the maximum principle addresses this difficulty. In place of (2.28) the necessary condition is that the Hamiltonian function $H[\mathbf{x}(t), \mathbf{u}(t), \boldsymbol{\lambda}(t), t]$ attains an absolute minimum as a function of $\mathbf{u}(t)$ over the admissible region Ω for all t in the interval (t_0, t_f). In other words, the optimal u_ε satisfies

$$H[\mathbf{x}(t), \mathbf{u}_\varepsilon(t), \lambda(t), t)] \le H[\mathbf{x}(t), \mathbf{u}(t), \boldsymbol{\lambda}(t), t] \qquad u_\varepsilon \in \Omega \tag{2.29}$$

Example 2.3. A linear differential system is described by

$$\dot{\mathbf{x}} = \mathbf{A}\mathbf{x} + \mathbf{B}\mathbf{u}$$

where

$$\mathbf{A} = \begin{bmatrix} 0 & 1 \\ 0 & 0 \end{bmatrix}, \qquad \mathbf{B} = \begin{bmatrix} 1 & 0 \\ 0 & 1 \end{bmatrix}, \qquad \mathbf{x}^T = [x_1, x_2], \qquad \mathbf{u}^T = [u_1, u_2]$$

find $u(t)$ such that

$$J = \tfrac{1}{2} \int_0^2 \|\mathbf{u}\|^2 \, dt$$

is minimum, given $\mathbf{x}^T(0) = [1, 1]$ and $x_1(2) = 0$

Define the Hamiltonian H as

$$H = \tfrac{1}{2}u^T u + \lambda^T(A\mathbf{x} + B\mathbf{u})$$

Equation (2.24) leads to the following costate equations:

$$\dot{\boldsymbol{\lambda}}(t) = -\frac{\partial H}{\partial \mathbf{x}}$$

$$\dot{\boldsymbol{\lambda}}(t) = A^T \boldsymbol{\lambda}$$

or

$$\dot{\lambda}_1(t) = 0 \tag{2.30a}$$

$$\dot{\lambda}_2(t) = -\lambda_1(t) \tag{2.30b}$$

with

$$\lambda_1(2) = \lambda_2(2) = 0 \qquad \text{(because } G = 0) \tag{2.30c}$$

Integration of equations (2.30a) and (2.30b) with the boundary (2.30c) gives

$$\lambda_1(t) = C_1 \tag{2.30d}$$

$$\lambda_2(t) = C_1(2 - t) \tag{2.30e}$$

H has a local minimum at

$$\frac{\partial H}{\partial \mathbf{u}(t)} = 0 \tag{2.30f}$$

or

$$\mathbf{u}(t) = -B^T \boldsymbol{\lambda}(t) \tag{2.30g}$$

we have

$$\dot{\mathbf{x}} = A\mathbf{x} + B\mathbf{u}$$

Substituting for $\mathbf{u}(t)$ from equation (2.30g) into the above equation, we obtain

$$\dot{\mathbf{x}} = A\mathbf{x} - BB^T \boldsymbol{\lambda} \tag{2.30h}$$

The solution of the above equation with equations (2.30d) and (2.30e) gives

$$x_1(t) = -C_1 t^2 + \tfrac{1}{6}C_1 t^3 + t - C_1 t + C_2$$

$$x_2(t) = -2C_1 t + \tfrac{1}{2}C_1 t^2 + C_3$$

Using the boundary conditions at $t = 0$, $\mathbf{x}^T(0) = [1, 1]$ gives

$$1 = C_2$$

$$1 = C_3$$

the state $x_1(t)$ now is given by

$$x_1(t) = -C_1 t^2 + \tfrac{1}{6}C_1 t^3 + t - C_1 t + 1$$

By using the boundary condition at $t = 2$, $x_1(2) = 0$

$$0 = -4C_1 + \tfrac{8}{6}C_1 + 2 - 2C_1 + 1$$

$$C_1 = \tfrac{9}{14}$$

We are now in a position to write the system equations which satisfied the boundary conditions as

$$\lambda_1(t) = \tfrac{9}{14}$$

$$\lambda_2(t) = \tfrac{9}{14}(2 - t)$$

$$x_1(t) = -\tfrac{9}{14}t^2 + \tfrac{3}{28}t^2 + \tfrac{5}{14}t + 1$$

$$x_2(t) = -\tfrac{9}{7}t + \tfrac{9}{28}t^2 + 1$$

$$u_1(t) = -\tfrac{9}{14}$$

$$u_2(t) = -\tfrac{9}{14}(2 - t)$$

Example 2.4. For the fixed plant dynamics given by $\dot{x}(t) = u(t)$, $x(0) = x_0$, determine the optimal closed-loop control which minimizes for fixed t_f

$$J = \tfrac{1}{2}Sx^2(t_f) + \tfrac{1}{2}\int_0^{t_f} u(t)^2\, dt$$

where S is an arbitrary constant. Do this by first determining the optimum open-loop control and trajectory and then let $u(t) = k(t)x(t)$.

Define the Hamiltonian as

$$H = \tfrac{1}{2}u^2(t) + \lambda(t)u(t)$$

We have

$$\dot{\lambda}(t) = -\frac{\partial H}{\partial x} = 0$$

Integrating directly we obtain

$$\lambda(t) = C_1 \tag{2.31a}$$

From the boundary term $G = \tfrac{1}{2}Sx^2(t_f)$, we have

$$\begin{aligned} \lambda(t_f) &= \left.\frac{\partial G}{\partial x(t)}\right|_{t=t_f} \\ \lambda(t_f) &= Sx(t_f) \end{aligned} \tag{2.31b}$$

Substituting from (2.31b) into (2.31a) at $t = t_f$, $\lambda(t_f) = Sx(t_f)$,

$$C_1 = Sx(t_f)$$
$$\lambda(t) = Sx(t_f) \tag{2.31c}$$

H has a local minimum at

$$\frac{\partial H}{\partial u(t)} = 0$$

or

$$0 = u(t) + \lambda(t)$$

Hence, the open-loop control is given by

$$u(t) = -Sx(t_f) \tag{2.31d}$$

Also, we have

$$\dot{x}(t) = u(t)$$

or

$$\dot{x}(t) = -Sx(t_f)$$

Integrating the above equation directly and by using at $t = 0$, $x(t) = x(0)$, we obtain

$$x(t) = -Sx(t_f)t + x(0) \tag{2.31e}$$

Now, the closed-loop control is given by

$$u(t) = k(t)[-Sx(t_f)t + x(0)]$$

or

$$u(t) = -Sk(t)x(t_f)t + k(t)x(0)$$

2.4. The Functional Analytic Optimization Technique

The aim of this section is to discuss the application of one important minimum norm result as an optimization technique which has been used as a powerful tool in the solution of problems treated in this book, especially problems stated in Chapters 4, 6, and 7. Before we do this, a brief discussion of relevant concepts from functional analysis is given (Ref. 2.5).

2.4.1. Norms

A norm, commonly denoted by $\|\cdot\|$, is a real-valued and positive definite scalar. The norm satisfies the following axioms:

(1) $\|x\| \geq 0$ for all $x \in X$, $\|x\| = 0 \leftrightarrow x = 0$

(2) $\|x + y\| \leq \|x\| + \|y\|$ for each $x, y \in X$

(3) $\|\alpha x\| = |\alpha| \cdot \|x\|$ for all scalars α and each $x \in X$

A normed linear (vector) space X is a linear space in which every vector x has a norm (length). The norm functional is used to define a distance and a convergence measure

$$d(x, y) = \|x - y\|$$

For example, let $[0, T]$ be a closed bounded interval. The space of continuous functions $x(t)$ on $[0, T]$ can have one of the following norms:

$$\|x\|_1 = \int_0^T |x(t)|\, dt$$

$$\|x\|_2 = \left[\int_0^T |x(t)|^2\, dt\right]^{1/2}$$

The normed linear space becomes a metric space under the (usual) metric (Ref. 2.4).

2.4.2. Inner Product (Dot Product) (Ref. 2.1)

Consider the real three-dimensional Euclidean space E^3. A vector or element in E^3 is an ordered real triple $x = (x_1, x_2, x_3)$ in which the norm is defined by

$$\|x\| = (|x_1|^2 + |x_2|^2 + |x_3|^2)^{1/2}$$

From linear algebra in a Euclidean space E if x is multiplied with another vector $y = (y_1, y_2, y_3)$, the result is (dot product)

$$\langle x, y\rangle = x_1y_1 + x_2y_2 + x_3y_3$$

or

$$\langle x, y\rangle = \sum_{i=1}^{3} x_iy_i$$

In the space E^3 the angle θ between x and y is also defined and is related to the norm by the equation

$$\langle x, y\rangle = \|x\| \cdot \|y\| \cos\theta$$

If the two vectors are orthogonal, $\theta = 90°$, then their inner product is zero

$$\langle x, y\rangle = 0$$

and they are collinear, $\theta = 0$, if their inner product is equal to

$$\langle x, y\rangle = \pm \|x\| \cdot \|y\|$$

In the complex space E^3 the inner product between any two vectors $x = (x_1, x_2, x_3)$ and $y(y_1, y_2, y_3)$ can be defined by

$$\langle x, y\rangle = x_1 y_1^* + x_2 y_2^* + x_3 y_3^*$$

Using these ideas as a background, let us now formulate the basic definition of an inner product in an abstract linear space.

Let X be a linear space. A rule that assigns a scalar $\langle x, y\rangle$ to every pair of elements $x, y \in X$ is called an inner product function if the following conditions are satisfied:

(1) $\langle x, y\rangle = \langle y, x\rangle$

(2) $\langle \alpha x + \beta y, z\rangle = \alpha\langle x, z\rangle + \beta\langle y, z\rangle$

(3) $\langle \lambda x, y\rangle = \lambda\langle x, y\rangle$

(4) $\langle x, y\rangle \geq 0, \qquad \langle x, x\rangle = 0 \leftrightarrow x = 0$

(5) $\langle x, x\rangle = \|x\|^2$

A linear space X is called a Hilbert space if X is an inner product space that is complete with respect to the norm induced by the inner product.

Equivalently, a Hilbert space is a Banach space whose norm is induced by an inner product. Let us now consider some specific examples of Hilbert spaces. The space E^n is a Hilbert space with inner product as defined by

$$\langle x, y\rangle = x^T y$$

or

$$\langle x, y\rangle = \sum_{i=1}^{n} x_i y_i$$

The space $L_2[0, T]$ is a Hilbert space with inner product

$$\langle x, y\rangle = \int_0^T x(t)y(t)\, dt$$

We have in this book a very useful Hilbert space. The elements of the space are vectors whose components are functions of time such as active power generation by the system units over the interval $[0, T_f]$. Given a positive definite matrix $B(t)$ whose elements are functions of time as well, we can define the Hilbert space $L_{2B}^n(0, T_f)$. The inner product in this case is given by

$$\langle V(t), U(t)\rangle = \int_0^{T_f} V^T(t)B(t)U(t)\,dt$$

for every $V(t)$ and $U(t)$ in the space.

2.4.3. Transformations

Let X and Y be linear vector spaces and let D be a subset of X. A rule that associates with every element $x \in D$ and element $y \in Y$ is said to be a transformation from X to Y with domain D. If y corresponds to x under T, we write $y = T(x)$ and y is referred to as the image of x under T. Alternatively, a transformation is referred to as an operator.

The transformation $T: X \to Y$ is said to be linear if

$$T(\alpha_1 x_1 + \alpha_2 x_2) = \alpha_1 T(x_1) + \alpha_2 T(x_2)$$

for all $\alpha_1, \alpha_2 \in R$ (the real line) and for every $x_1, x_2 \in X$.

The linear operator T from a normed space X to a normed space Y is said to be bounded if there is a constant M such that $\|Tx\| \leq M\|x\|$ for all $x \in X$. The normed space of all bounded linear operators from the normed space X into the normed space Y is denoted by $B(X, Y)$. Examples of bounded linear operators include one transformation useful for our purposes. This is $T: L_{2B}^n(0, T_f) \to R^m$ defined by $b = T[U(t)]$

$$b = \int_0^{T_f} M^T U(t)\,dt$$

In practical applications there are many transformations; among these transformations we have (Ref. 2.5) the following:

(1) If two power plants supply a demand $P_D(t)$ such that the active power balance equation is satisfied

$$P_D(t) = P_1(t) + P_2(t)$$

If time is not included as a parameter the above equation can be written as

$$P_D(t) = M^T P(t)$$

where

$$M = \text{col}(1, 1)$$

$$P(t) = \text{col}[P_1(t), P_2(t)]$$

This defines a transformation $T: L_2^2(0, T_f) \to L_2(0, T_f)$ sending functions $[P_1(t), P_2(t)]$ into their image $P_D(t)$. Observe that $T = M^T$.

A functional is a transformation from a linear space into the space of real or complex scalars. A typical functional is the objective functional of optimal economy operation of m thermal plants given by

$$J(P_s) = \int_0^{T_f} \sum_{i=1}^{m} [\alpha_i + \beta_i P_{si}(t) + \gamma_i P_{si}^2(t)]\, dt$$

In the above equation the space $L_2^m(0, T_f)$ of thermal power generation vector functions is mapped into the real scalar space. In a linear space X, a functional f is linear if for any two vectors $x_1, x_2 \in X$, and any two scalars α_1 and α_2 we have

$$f(\alpha_1 x_1 + \alpha_2 x_2) = \alpha_1 f(x_1) + \alpha_2 f(x_2)$$

On a normed space a linear functional f is bounded if there is a constant M such that $|f(x)| \leq M\|x\|$ for every $x \in X$. The space of these linear functionals is a normed linear space X^*. X^* is the normed dual space of X and is a Banach space. If X is a Hilbert space then $X = X^*$. Thus Hilbert spaces are self-dual. For the normed linear space to be a reflexive $X = X^{**}$. Any Hilbert space is reflexive.

Let X and Y be normed spaces and let $T \in B(X, Y)$. The adjoint (conjugate) operator $T^*: Y^* \to X^*$ is defined by

$$\langle x, T^* y\rangle = \langle Tx, y^*\rangle$$

An important special case is that of a linear operator $T: H \to G$ where H and G are Hilbert spaces. If G and H are real, then they are their own duals and the operator T^* can be regarded as mapping G into H. In this case the adjoint relation becomes

$$\langle Tx, y\rangle = \langle x, T^* y\rangle$$

Note that the left-hand side inner product is taken in G while the right-hand side inner product is taken in H.

Composite transformations can be formed as follows. Suppose T and G are transformations $T: X \to Y$ and $G: Y \to Z$. We define the transformation $GT: X \to Z$ by

$$(GT)(x) = G(T(x))$$

We then say that GT is a composite of G and T, respectively.

2.4.4. A Minimum Norm Theorem (Ref. 2.5)

With the above-outlined definitions in mind, we can now introduce one powerful result in optimization theory. The theorem described here is only one of a wealth of results that utilize functional analytic concepts to effectively solve optimization problems.

Theorem. Let B and D be Banach spaces. Let T be a bounded linear transformation defined on B with values in D. Let $\hat{u}$ be a given vector in B. For each ξ in the range of T, there exists a unique element $u_\xi \in B$ that satisfies

$$\xi = Tu$$

while minimizing the objective functional

$$J(u) = \|u - \hat{u}\|$$

The unique optimal $u_\xi \in B$ is given by

$$u_\xi = T^+[\xi - T\hat{u}] + \hat{u}$$

where the pseudoinverse operator T^+ in the case of Hilbert spaces is given by

$$T^+\xi = T^*[TT^*]^{-1}\xi$$

provided that the inverse of TT^* exists.

The theorem as stated is an extension of the fundamental minimum norm problem where the objective functional is

$$J(u) = \|u\|$$

The optimal solution for this case is

$$u_\xi = T^+\xi$$

with T^+ being the pseudoinverse associated with T.

The above equations for the optimal control vector u can be obtained by using the Lagrange multiplier argument. The augmented cost functional can be obtained by adjoining to the cost functional the equality constraint via Lagrange's multiplier as follows:

$$\tilde{J}(u) = \|u - \hat{u}\|^2 + \langle \lambda, (\xi - Tu)\rangle$$

where λ is a multiplier (in fact $\lambda \in D$) to be determined so that the constraint $\xi = Tu$ is satisfied. By utilizing properties of the inner product we can write

$$\tilde{J}(u) = \|u - \hat{u} - T^*(\lambda/2)\|^2 - \|T^*(\lambda/2)\|^2 + \langle \lambda, \xi\rangle$$

Only the first norm of the above equation depends explicitly on the control vector u. To minimize $\tilde{J}$ we consider only

$$\tilde{J}(u) = \|u - \hat{u} - T^*(\lambda/2)\|$$

The minimum of the above equation is clearly achieved when

$$u_\xi = \hat{u} + T^*(\lambda/2)$$

To find the value of $(\lambda/2)$ we use the equality constraint

$$\xi = Tu_\xi$$

which gives

$$(\lambda/2) = [TT^*]^{-1}[\xi - T\hat{u}]$$

It is therefore clear that with an invertible TT^* we write

$$u_\xi = T^*[TT^*]^{-1}[\xi - T\hat{u}]$$

which is the required result. In the above equation if $\hat{u} = 0$ we obtain

$$u_\xi = T^*[TT^*]^{-1}\xi$$

which is the same result obtained for the fundamental minimum norm problem.

In applying this result to our physical problem we need to recall two important concepts from ordinary constrained optimization. These are the Lagrange multiplier rule and the Kuhn–Tucker multipliers. An augmented objective functional is obtained by adjoining to the original cost functional terms corresponding to the constraints using the necessary multipliers. The object in this case is to ensure that the augmented functional can indeed be cast as a norm in the chosen space. A set of linear constraints on the control vector is singled out; this set, under appropriate conditions, will define the bounded linear transformation T.

References

2.1. Porter, W. A., *Modern Foundations of Systems Engineering*, Macmillan, New York, 1966.

2.2. Luenberger, D. G., *Optimization by Vector Space Methods*, Wiley, New York, 1969.

2.3. Sage, A., *Optimum System Controls*, Prentice-Hall, Englewood Cliffs, New Jersey, 1968.

2.4. Sage, A. P., and White, C. C., *Optimum Systems Control*, Prentice-Hall, Englewood Cliffs, New Jersey, 1977.

2.5. El-Hawary, M. E., and Christensen, G. S., *Optimal Economic Operation of Electric Power Systems*, Academic, New York, 1979.

2.6. Bryson, A. E., and Ho, Y. C., *Applied Optimal Control*, Wiley, New York, 1975.

2.7. Rao, S. S., *Optimization Theory and Applications*, Wiley Eastern Limited, New Delhi, India, 1979.

2.8. Leitmann, G., *The Calculus of Variations and Optimal Control*, Plenum Press, New York, 1981.

2.9. Kirk, D. E., *Optimal Control Theory: An Introduction*, Prentice-Hall, Englewood Cliffs, New Jersey, 1970.

2.10. Narici, B., *Functional Analysis*, Academic, New York, 1966.

3

Dynamic Optimal Load Flow

3.1. Introduction

The load flow problem in an electric power system is concerned with solving a set of static nonlinear equations describing the electric network performance. The problem is formulated on the basis of Kirchhoff's laws in terms of active and reactive power injections and voltages at each node in the system. Presently, the load flow program is a major tool for the electric power systems engineer in carrying out diverse functions in the planning and operation of the system. This topic is treated in detail in many introductory electric power systems textbooks (Refs. 3.1–3.4).

In a load flow problem, the active power and voltage magnitudes are specified values. For load buses, the active and reactive power demands are also given. One generator bus is taken as reference (slack bus) with specified voltage magnitude and phase angle. The problem is to find the reactive power generation and phase angles at the generating buses along with voltage magnitudes and their angles at the load buses. Available software uses the powerful Newton's iterative method to obtain the solution.

For electric power systems with thermal (steam) generating resources, an important economic operation function is that of optimal load flow (OLF). In contrast to the load flow problem, where active power generations are specified, in an OLF, the optimal generations are sought to minimize the operating cost of the system.

Optimal load or power flow (OPF) has been the subject of continuous intensive research and algorithmic improvements since its introduction in the early 1960s by Squires (Ref. 3.5) and Carpentier (Ref. 3.6). The OPF problem seeks to find an optimal profile of active and reactive generations along with voltage magnitudes in such a manner as to minimize the total operating costs of an all-thermal electric power system, while satisfying network security constraints. The problem is one of static optimization involving a large-scale system. Practical OPF algorithms for energy control center implementation have been developed and reported in 1984 by Sun *et al.* (Ref. 3.7) and Burchett *et al.* (Ref. 3.8). The results of a detailed study assessing the benefits to the Ontario Hydro System due to OPF implementation were reported in 1985 by El-Kady *et al.* (Ref. 3.9). OPF is a formulation that is more realistic than the conventional economic dispatch function, which does not account for network security constraints.

This chapter deals with the more complex problem of hydrothermal optimal power flow (HTOPF), where limited-energy hydrogeneration resources are present in the system. A major difference between the HTOPF and OPF problems is that the former is a dynamic problem where decisions at an early time in the optimization interval influence decisions at later times. Conventional economic scheduling of hydrothermal systems is concerned with a system represented by an active power balance model. This area was pioneered by Ricard (Ref. 3.10), Chandler *et al.* (Ref. 3.11), and Glimm and Kirchmayer (Ref. 3.12).

The earliest work on optimal load flow for hydrothermal systems was probably due to Ramamoorthy and Gopala Rao (Ref. 3.13). Head variation was neglected and nonlinear programming techniques were employed. The minimum norm formulation of an optimal load flow for hydrothermal systems with isolated hydroplants was given in El-Hawary and Christensen (Ref. 3.14) in 1973. Various other methods for tackling this problem are discussed in a book by the same authors (Ref. 3.15). The area of economic operation of electric power systems is explored fully from a tutorial point of view in Refs. 3.15–3.17.

We present the formulation and optimality conditions for systems with fixed-head hydroplants subject to an energy-limited constraint. The solution to obtain the optimal operational strategy is implemented using Newton's method. Because of the large-scale nature of the problem, special sparsity oriented and structural property enhancements are explored and four versions for algorithmic implementation are detailed.

Special attention is paid to finding realistic initial guess estimates for the algorithms.

3.2. The All-Thermal OPF

In an all-thermal power system, the total production cost in simple terms is considered to be the total fuel cost over the interval of consideration. We express this as

$$J = \int_0^{T_f} \sum_{i \in R_s} F_i(P_{gi})\, dt \tag{3.1}$$

where R_s is the set of thermal units and $F_i(P_{gi})$ is the fuel cost of the ith thermal unit, which is a function of the level of power generation at the ith unit.

The associated constraints relate to the balance of active and reactive power in the network. The total power generation with or without losses included must meet and load demand, which is assumed known a priori. A number of models to describe the network performance have been proposed. Two of these models are as follows:

1. Active Power Balance Equation:
   a. Without losses

$$\sum_{i \in R_s} P_{gi} - P_D = 0 \tag{3.2}$$

   where P_{gi} is the ith unit generation and P_D is the total load demand.

   b. With losses

$$\sum_{i \in R_s} P_{gi} - [P_L(P_g) + P_D] = 0 \tag{3.3}$$

   where $P_L(P_g)$ is the power loss and is considered a function of the power generation only.

2. Load Flow Equations:
   In an N-node network we have

$$P_i(V, \delta) - P_{gi} + P_{di} = 0, \qquad i = 1, \ldots, N \tag{3.4a}$$

$$Q_i(V, \delta) - Q_{gi} + Q_{di} = 0, \qquad i = M + 1, \ldots, N \tag{3.4b}$$

where M is the number of voltage-controlled buses

$$V_i = V_i \underline{/\delta_i} \tag{3.5a}$$

$$Y_{ij} = Y_{ij} \underline{/\psi_{ij}} \tag{3.5b}$$

$$P_i = V_i \sum_{i=1}^{N} Y_{ij} V_j \cos(\delta_i - \delta_j - \psi_{ij}) \tag{3.5c}$$

$$Q_i = V_i \sum_{i=1}^{N} Y_{ij} V_j \sin(\delta_i - \delta_j - \psi_{ij}) \tag{3.5d}$$

where P_i is active power injection into bus i, Q_i is the reactive power injection into bus i, P_{di} and Q_{di} are the active and reactive loads on bus i, and Y_{ij} and ψ_{ij} are the magnitude and angle of the admittance matrix elements.

The dimensionality of the problem will be much larger if we incorporate the load flow equations into the formation; however, the losses are more accurate and the actual state of the system appears as a by-product of the optimal solution. In our present formulation, we employ the exact load flow equations as the network model.

This constrained minimization problem can be transformed into an unconstrained minimization problem by augmenting the load flow constraints into the objective function. The additional variables are commonly known as Lagrange multiplier functions or incremental cost functions in power system terminology. The new cost functional becomes

$$J_1 = \int_0^{T_f} \left\{ \sum_{i \in R_s} F_i(P_{gi}) + \sum_{i=1}^{N} \lambda_{pi}[P_i(V, \delta) - P_{gi} + P_{di}] \right.$$
$$\left. + \sum_{i=M+1}^{N} \lambda_{qi}[Q_i(V, \delta) - Q_{gi} + Q_{di}] \right\} dt \tag{3.6}$$

The integrand is a function of time and the form of the multiplier functions is still an unknown. Our task is to determine the expression of the integrand at different time intervals such that the cost functional J_1 is minimized. In other words, if we can determine at each time interval the power generation levels, the voltage magnitudes, the phase angles, and the multiplier λ's, the problem will then be solved. Making use of variational principles, we can derive a set of equations that can then be solved for the unknowns of the problem. The set of equations is obtained by taking the first derivatives of the integrand with respect to the control variables. Therefore, at each time interval, we have a system of nonlinear equations:

$$\frac{\partial F_i}{\partial P_{gi}} - \lambda_{Pi} = 0, \qquad i \in R_s \tag{3.7a}$$

$$\sum_{i=1}^{N} \lambda_{pi} \frac{\partial P_i(V, \delta)}{\partial \delta_i} + \sum_{i=M+1}^{N} \lambda_{qi} \frac{\partial Q_i(V, \delta)}{\partial \delta_j} = 0, \qquad j = 2, \ldots, N \tag{3.7b}$$

$$\sum_{i=1}^{N} \lambda_{pi} \frac{\partial P_i(V, \delta)}{\partial V_j} + \sum_{i=M+1}^{N} \lambda_{qi} \frac{\partial Q_i(V, \delta)}{\partial V_j} = 0, \qquad j = M+1, \ldots, N \tag{3.7c}$$

$$P_i(V, \delta) - P_{gi} + P_{di} = 0, \qquad i = 1, \ldots, N \tag{3.7d}$$

$$Q_i(V, \delta) - Q_{gi} + Q_{di} = 0, \qquad i = M+1, \ldots, N \tag{3.7e}$$

The number of unknowns in the above is equal to the number of equations. Hence, the solution is well defined if it exists. The solution for each time interval can be obtained independently because there is no coupling between the variables in different time intervals. The problem, therefore, for a system with thermal resources is the well-known optimal load flow.

3.3. The Hydrothermal OPF

There is no fuel cost associated with the operation of hydraulic units. The input–output characteristic of a hydrogenerator is represented by the variation of water discharge q as a function of power output P_g and net head h. We write the Glimm–Kirchmayer model (Ref. 3.12)

$$q_i = K\psi(h)\phi(P_{gi}), \qquad i \in R_H \qquad \text{(MCF/hr)} \tag{3.8a}$$

where R_H is the set of hydrogenerators, K is a constant, and ψ and ϕ are two independent functions. In the case of constant head, $\psi(h)$ becomes a constant, and the above equation is rewritten as

$$q_i = K'\phi(P_{gi}), \qquad i \in R_H \tag{3.8b}$$

where K' is a new constant formed by the multiplication of K and $\psi(h)$.

We include the hydraulic units in the cost functional of equation (3.6) by using the constant multipliers ν_i (\$/MCF), which accounts for the unit conversion from MCF/hr to \$/hr. This constant multiplier is more commonly known as the water conversion factor or water-worth. The new cost functional becomes

$$J_2 = \int_0^{T_f} \left\{ \sum_{i\in R_s} F_i(P_{gi}) + \sum_{i\in R_H} \nu_i q_i + \sum_{i=1}^{N} \lambda_{Pi}[P_i(V,\delta) - P_{gi} + P_{di}] \right.$$
$$\left. + \sum_{i=M+1}^{N} \lambda_{qi}[Q_i(V,\delta) - Q_{gi} + Q_{di}] \right\} dt - \sum_{i\in R_H} \nu_i b_i \tag{3.9}$$

It is assumed that the sets R_s and R_k are defined a priori as the outcome of a unit commitment schedule for the system over the optimization interval.

The optimality conditions with respect to the additional variables $P_{gi}(i \in R_H)$ result in

$$\nu_i \frac{\partial q_i}{\partial P_{gi}} - \lambda_{Pi} = 0, \qquad i \in R_H \tag{3.10a}$$

The new constant multipliers ν_i are evaluated by incorporating the water volume constraints:

$$\int_0^{T_f} q_i(t)\, dt = b_i, \qquad i \in R_H \tag{3.10b}$$

where b_i is a prespecified amount of water available for the ith hydraulic unit during the optimization interval considered.

The set of equations (3.7) and (3.10) constitute the necessary requirements for the optimal solution of out HTOPF problem. The set of optimality equations assuming that the fuel cost and water discharge models are quadratic is given by

$$f_{si}(t):\quad (\beta_i + 2\gamma_i P_{gi}) - \lambda_{Pi} = 0, \qquad i = 1, \ldots, N_s \tag{3.11a}$$

$$f_{hi}(t):\quad \nu_i(\beta_i + 2\gamma_i P_{gi}) - \lambda_{Pi} = 0, \qquad i = N_s + 1, \ldots, N_g \tag{3.11b}$$

$$f_{\delta j}(t):\quad \sum_{i=1}^{N} \lambda_{Pi} \frac{\partial P_i(V, \delta)}{\partial \delta_j} + \sum_{i=M+1}^{N} \lambda_{qi} \frac{\partial Q_i(V, \delta)}{\partial \delta_j} = 0, \qquad j = 2, \ldots, N \tag{3.11c}$$

$$f_{vj}(t):\quad \sum_{i=1}^{N} \lambda_{Pi} \frac{\partial P_i(V, \delta)}{\partial V_j} + \sum_{i=M+1}^{N} \lambda_{qi} \frac{\partial Q_i(V, \delta)}{\partial V_j} = 0,$$

$$j = M + 1, \ldots, N \tag{3.11d}$$

$$f_{Pi}(t):\quad P_i(V, \delta) - P_{gi} + P_{di} = 0, \qquad i = 1, \ldots, N \tag{3.11e}$$

$$f_{qi}(t):\quad Q_i(V, \delta) - Q_{gi} + Q_{di} = 0, \qquad i = M + 1, \ldots, N \tag{3.11f}$$

$$f_{\nu i}:\quad \int_0^{T_f} (\alpha_i + \beta_i P_{gi} + \gamma_i P_{gi}^2)\, dt - b_i = 0, \qquad i = N_s + 1, \ldots, N_g \tag{3.11g}$$

The major difference between the problem for an all-thermal system and that for a hydrothermal system is the time-coupling effect. The former can be solved independently for each time interval. The latter is restricted to considering all time intervals as a whole owing to the volume of water constraint. The dimensionality of the problem increases as the number of time intervals considered increases. Typically, a daily optimization process is considered on an hourly basis, which means that there are 24 discrete time intervals. For a 5-bus system with one thermal and one hydrogenerator, the number of variables in each time interval is 17, which is calculated from equation (3.12). Therefore, the dimension of the problem for 3 discrete time intervals is 52 and 24 discrete time intervals becomes 409. It is assumed that a stepwise approximation to the time functions involved is employed in the solution algorithms:

Number of variables per time interval

$$= N_g + (N - 1) + (N - M) + N + (N - M)$$

$$= Ng + 4N - 2M - 1 \tag{3.12}$$

where N_g is the number of generator buses in the system. In addition there are N_h unknown water conversion coefficients.

In the practical problem there are some other security-related constraints called soft constraints, which indicate the desired limits. Among these, we have the variable constraint of the form

$$x_i^{\min} \leq x_i \leq x_i^{\max}$$

where x_i can be a physical quantity such as bus voltage magnitude, phase angle difference, or active power generation. The other type, called functional constraint, is of the form

$$g_i^{\min} \leq g_i(X) \leq g_i^{\max}$$

where g_i is a physical quantity that is a function of the control variables, and it can represent a quantity such as reactive power generation, active power flow of line, or reactive power flow of line. All the soft constraints mentioned can be accounted for by a penalty function approach or they can be checked during the solution process so that none of the limits are violated in accordance with the Kuhn–Tucker approach.

Inequality constraints on the problem variables considered are

1. Upper and lower bounds on the voltage magnitudes at the load buses.
2. Upper and lower bounds on the active and reactive power generation at generator buses.
3. Upper and lower bounds on the reactive power injection at buses with synchronous condensers.

The Kuhn–Tucker approach is adopted and thus inequality constraints are assumed to be inactive during the initial solution process. The result is checked for compliance with the inequality constraint requirements. If there is a variable violation at the ith bus, then the variable is set at the limiting value and the bus type is changed accordingly. If there is a Q violation at a PV bus, it is then changed to a PQ bus with corresponding voltage constraints and the solution process is repeated until there are no limit violations at the optimum. In this case, the converged solution would be considered as the optimal solution to the constrained problem since the Kuhn–Tucker conditions are satisfied.

3.4. A Successive Approximation Solution

The hydrothermal optimal load flow problem can be solved using a successive approximation algorithm developed by El-Hawary, Rao, and Christensen (Ref. 3.18). In this approach, the active power equation (3.4a) for the slack bus ($i = 1$) is not incorporated directly into the optimization process. Instead we substitute for P_1 in the objective functional (3.6) using

$$P_{g1} = P_1(V, \delta) + P_{d1} \tag{3.13}$$

As a result a set of optimality conditions which is an alternative to (3.11) is obtained.

The concrete form of optimality conditions is given as follows: At the thermal generation buses, we have

$$\beta_i + 2\gamma_i P_{gi} - \lambda_{pi} - 0, \qquad i = 2, \ldots, N_s \tag{3.14}$$

At the hydrogeneration buses, we have

$$\nu_i(\beta_i + 2\gamma_i P_{gi}) - \lambda_{pi} = 0, \qquad i = N_s + 1, \ldots, N_g \tag{3.15}$$

At all the system buses with the exception of the slack bus, we have

$$\beta_1 \frac{\partial P_1(V, \delta)}{\partial \delta_j} + 2\gamma_1[P_1(V, \delta) + P_{di}] \frac{\partial P_1(V, \delta)}{\partial \delta_j} + \sum_{i=2}^{N} \lambda_{pi} \frac{\partial P_i(V, \delta)}{\partial \delta_j} + \sum_{i=M+1}^{N} \lambda_{qi} \frac{\partial Q_i(V, \delta)}{\partial \delta_j} = 0, \qquad j = 2, \ldots, N \tag{3.16}$$

At the load buses, we have

$$\beta_1 \frac{\partial P_1(V, \delta)}{\partial |V_j|} + 2\gamma_1[P_1(V, \delta) + P_{d1}] \frac{\partial P_1(V, \delta)}{\partial |V_j|} + \sum_{i=2}^{N} \lambda_{pi} \frac{\partial P_i(V, \delta)}{\partial |V_j|} + \sum_{i=M+1}^{N} \lambda_{qi} \frac{\partial Q_i(V, \delta)}{\partial |V_j|} = 0, \qquad J = M + 1, \ldots, N \tag{3.17}$$

In addition we have to satisfy the load flow equations (3.4) for each time instant. For the overall interval $[0, T_f]$ we have one integral constraint for each hydroplant given by the value of the water constraint

$$\int_0^{T_f} (\alpha_i + \beta_i P_{gi} + \gamma_i P_{gi}^2)\, dt = b_i, \qquad i = N_s + 1, \ldots, N_g \tag{3.18}$$

Equations (3.16) and (3.17) can be cast into a more compact form by defining the following variables:

$$f_{\delta j}(t) = \beta_1 \frac{\partial P_1(V, \delta)}{\partial \delta_j} + 2\gamma_1[P_1(V, \delta) + P_{d1}] \frac{\partial P_1(V, \delta)}{\partial \delta_j}, \qquad j = 2, \ldots, N$$

$$f_{|V_j|}(t) = \beta_1 \frac{\partial P_1(V, \delta)}{\partial |V_j|} + 2\gamma_1[P_1(V, \delta) + P_{d1}] \frac{\partial P_1(V, \delta)}{\partial |V_j|}, \qquad j = M + 1, \ldots, N$$

$$J_{p\delta_{ij}}(t) = \frac{\partial P_i(V, \delta)}{\partial \delta_j}, \qquad i = 2, \ldots, N, \qquad j = 2, \ldots, N$$

$$J_{Q\delta_{ij}}(t) = \frac{\partial Q_i(V, \delta)}{\partial \delta_j}, \qquad i = M+1, \ldots, N, \qquad j = 2, \ldots, N$$

$$J_{PV_{ij}}(t) = \frac{\partial P_i(V, \delta)}{\partial |V_j|}, \qquad i = 2, \ldots, N, \qquad j = M+1, \ldots, N$$

$$J_{QV_{ij}}(t) = \frac{\partial Q_i(V, \delta)}{\partial |V_j|}, \qquad i = M+1, \ldots, N, \qquad j = M+1, \ldots, N$$

As a result equations (3.16) and (3.17) are rewritten as

$$f_{\delta j}(t) + \sum_{i=2}^{N} \lambda_{pi} J_{P\delta_{ij}} + \sum_{i=M+1}^{N} \lambda_{qi} J_{Q\delta_{ij}} = 0, \qquad j = 2, \ldots, N \tag{3.19}$$

$$f_{|V_j|}(t) + \sum_{i=2}^{N} \lambda_{pi} J_{PV_{ij}} + \sum_{i=M+1}^{N} \lambda_{qi} J_{QV_{ij}} = 0, \qquad j = M+1, \ldots, N \tag{3.20}$$

In vector form we write

$$\begin{array}{c} \\ N-1 \\ \\ N-M \end{array} \overset{\begin{array}{cc} \leftarrow N-1 \rightarrow & \leftarrow N-M \rightarrow \end{array}}{\left[\begin{array}{c|c} J_{P\delta_{ji}} & J_{Q\delta_{ji}} \\ \hline J_{PV_{ji}} & J_{QV_{ji}} \end{array}\right]} \overset{\leftarrow 1 \rightarrow}{\left[\begin{array}{c} \lambda_{pi} \\ \hline \lambda_{qi} \end{array}\right]} = - \overset{\leftarrow 1 \rightarrow}{\left[\begin{array}{c} f_{\delta j} \\ \hline f_{|V_j|} \end{array}\right]} \begin{array}{c} N-1 \\ \\ N-M \end{array} \tag{3.21}$$

or

$$\hat{J}^T \lambda = -\hat{f} \tag{3.22}$$

Note that $\hat{J}^T$ is the transpose of the Jacobian of the load flow equations.

The optimality equations (3.14), (3.15), and (3.21) together with the constraint equations (3.4) and (3.18) completely specify the optimal operational strategy for the system, assuming no inequality constraints on the reactive power generation at the voltage-controlled buses. Solving these equations yields the value of P_{gi} $(i = 2, \ldots, N_g)$ for optimal operation. The values of P_1 and Q_i $(i = 1, \ldots, M)$ can be calculated as a by-product. The solution technique is discussed next.

Let us now consider the modification to the equations, if the inequality constraints on some of the Q_i are violated. Suppose that this happens at m of the M voltage-controlled buses at some instant; as mentioned earlier, Q_i are set to the violated limit ($Q_{\min i}$ or $Q_{\max i}$, as the case may be) at those buses and the voltages at those buses become variables. This means that we have m more reactive power flow equations, and m more variables, at the instant under consideration. Our set of reactive power flow equations is written as

$$Q_i(V, \delta) - Q_{gi} + Q_{di} = 0, \qquad i = M - m + 1, \ldots, N \tag{3.23}$$

Also, equation (3.20) is now written as

$$\begin{array}{c} \\ \uparrow N-1 \downarrow \\ \uparrow N-M+m \downarrow \end{array} \begin{array}{c} \leftarrow N-1 \rightarrow \quad \leftarrow N-M+m \rightarrow \\ \left[\begin{array}{c|c} J_{P\delta_{ji}} & J_{Q\delta_{ji}} \\ \hline J_{PV_{ji}} & J_{QV_{ji}} \end{array}\right] \end{array} \begin{array}{c} \leftarrow 1 \rightarrow \\ \left[\begin{array}{c} \lambda_{pi} \\ \hline \lambda_{qi} \end{array}\right] \end{array} = - \begin{array}{c} \leftarrow 1 \rightarrow \\ \left[\begin{array}{c} f_{\delta_j} \\ \hline f_{|V_j|} \end{array}\right] \end{array} \begin{array}{c} \\ \uparrow N-1 \downarrow \\ \uparrow N-M+m \downarrow \end{array} \tag{3.24}$$

or alternatively

$$\hat{J}_m^T \lambda = -\hat{f}_m \tag{3.25}$$

where $\hat{J}_m^T$ is the transpose of the Jacobian of the load flow equations (3.4) and (3.23). The remaining equations are unaltered. We are now in a position to solve for $P_{gi}(i = 2, \ldots, N_g)$ using the optimal equations (3.14), (3.15), and (3.25) and the constraint equations (3.4), (3.23), and (3.18). Note that the inequality constraints on Q_i have now been automatically taken into account.

The optimization procedure involves a two-step process. In the first step we optimize the generating schedule without taking into account the inequality constraints on Q_i and check whether the limits on Q_i are violated. If at some instant the limits are violated at m of the M buses, we replace Q_i by the violated values and recompute P_{gi} using the augmented set of equations for that instant. Note that violation may occur at different set of buses at different instants.

Equations (3.14) and (3.15) yield the following expressions for P_{gi}:

$$P_{gi} = \frac{1}{2_{\gamma i}}[\lambda_{pi} - \beta_i], \qquad i = 2, \ldots, N_s \tag{3.26}$$

$$P_{gi} = \frac{1}{2_{\gamma i}}\left[\frac{\lambda_{pi}}{\cdot \nu_i} - \beta_i\right], \qquad i = N_s + 1, \ldots, N_g \tag{3.27}$$

Substituting for P_{gi} from (3.26) into (3.18) and simplifying, the volume of water constraint becomes

$$\int_0^{T_f} \left[\alpha_i - \frac{\beta_i^2}{4_{\gamma i}} + \frac{\lambda_{pi}^2}{4\gamma_i \nu_i^2}\right] dt = b_i \tag{3.28}$$

Integrating (3.28) and putting $\int_0^{T_f} \lambda_{pi}^2 \, dt = \Lambda_{pi}$ we get

$$\left[\alpha_i - \frac{\beta_i^2}{4_{\gamma i}}\right] T_f + \frac{\Lambda_{pi}}{4\gamma_i \nu_i^2} = b_i$$

from which

$$\nu_1 = \Lambda_{pi}^{1/2}[4\gamma_i(b_i - \alpha_i T_f) + \beta_i^2 T_f]^{-1/2} \tag{3.29}$$

Substituting (3.29) into (3.27), we obtain

$$P_{gi} = \frac{1}{2_{\nu i}}\{\lambda_{pi}\Lambda_{pi}^{-1/2}[4\gamma_i(b_i - \alpha_i T_f) + \beta_i^2 T_f]^{1/2} - \beta_i\}, \qquad i = N_{s+1}, \ldots, N_g \tag{3.30}$$

Procedure

We are now in a position to formulate the solution algorithm, which proceeds as follows:

1. Choose initial guess controls P_{gi}^0 $(i = 2, \ldots, N_g)$.
2. Solve the load flow equations (3.4) thus obtaining the initial guess $|V_1|^0$ $(i = M + 1, \ldots, N)$, δ_i^0 $(i = 2, \ldots, N)$.
3. Compute the load flow Jacobian J at the solution points $|V_i|^0$ $(i = M + 1, \ldots, N)$, δ_i^0 $(i = 2, \ldots, N)$.
4. Solve (3.21) for λ_{pi} $(i = 2, \ldots, N)$, λ_{qi} $(i = M + 1, \ldots, N)$.
5. Substitute λ_{pi} in (3.26) to get a new set of P_{gi}^n $(i = 2, \ldots, N_s)$ for the thermal plants.
6. Find Λ_{pi} by integrating λ_{pi}^2 over the optimization interval, $i = N_s + 1, \ldots, N_g$.
7. Substitute λ_{pi} and Λ_{pi} in (3.30) to get a new set of P_{gi}^n $(i = N_s + 1, \ldots, N_g)$ for the hydroplants.
8. Compare the P_{gi}^n's with the initial guess values P_{gi}^0; if they are close the optimal generations P_{gi} $(i = 2, \ldots, N_g)$ (ignoring the inequality constraints) are obtained; otherwise an updated, hopefully improved, control vector P_{gi}^u is chosen for the initial guess, and steps 1–8 repeated until convergence is obtained.
9. Compute P_{g1} and Q_{gi} $(i = 1, \ldots, M)$.
10. Check whether the inequality constraints are violated. If not, the converged P_{gi} $(i = 2, \ldots, N_g)$ together with the calculated P_{g1} give the optimal generation schedule.
11. If the inequality constraints are violated at m of the M control buses at any instant, set Q_{gi} to the violated limits at that instant.
12. Start again with the initial guess vector P_{gi}^0 $(i = 2, \ldots, N_g)$.
13. Solve the augmented load flow equations (3.4) to obtain $|V_i|^0$ $(i = M - m + 1, \ldots, N)$, δ_i^0 $(i = 2, \ldots, N)$.

14. Compute the augmented load flow Jacobian J_m at the solution points $|V_i|^0$ $(i = M - m + 1, \ldots, N)$, δ_i^0 $(i = 2, \ldots, N)$.

15. Solve (3.24) for λ_{pi} $(i = 2, \ldots, N)$, λ_{qi} $(i = M - m + 1, \ldots, N)$.

16. Repeat steps 5–7 to obtain a new set of P_{gi}^n $(i = 2, \ldots, N_g)$.

17. Compare the new P_{gi}^n with the initial guess values P_{gi}^0; if close, the optimal P_{gi} $(i = 2, \ldots, N_g)$ including the inequality constraints, are obtained. If not, an updated P_{gi}^u is chosen for the initial guess, and steps 12–17 are repeated until convergence is obtained. P_{g1} can now be computed, and we have the complete optimal generation schedule taking into account the inequality constraints on reactive power generation.

Choice of Initial Guess Vector P_{gi}^0. This choice should be based on some realistic estimates. The volume of water constraint dictates the maximum possible generation at the hydrobuses. Assuming that P_{gi}^0 remains constant over the optimization interval at each of the hydrobuses, we can use equation (3.30) to find the value of P_{gi}^0 at those buses. Note that if P_{gi} is constant, λ_{pi} is constant, and $\Lambda_{pi} = T_f \lambda_{pi}^2$. The expression for P_{gi}^0 then reduces to

$$P_{gi}^0 = \frac{1}{2_{\gamma i}} \left\{ \left[4\gamma_i \left(\frac{b_i}{T_f} - \alpha_i \right) + \beta_i^2 \right]^{1/2} - \beta_i \right\} \tag{3.31}$$

Equation (3.31) can be used to obtain the initial guess P_{gi}^0 $(i = N_s + 1, \ldots, N_g)$ for the hydrobuses.

We now proceed to estimate the initial guess value of P_{gi}^0 for the thermal buses. Neglecting line losses, the total active power demand to be met by the generators at the thermal buses is given by

$$P_D = \sum_{i=1}^{N} P_{di} - \sum_{i=N_s+1}^{N_g} P_{gi} \tag{3.32}$$

Further the optimum generation at N_s thermal buses belonging to a set R_s, to satisfy a given active power demand P_d, under negligible losses, is given by

$$P_{si\xi} = (\lambda - \beta_i)/2\gamma_i, \qquad i \in R_S \tag{3.33}$$

where

$$\lambda = \left[2P_D + \sum_{i \in R_s} (\beta_i/\gamma_i) \right] \Big/ \left(\sum_{i \in R_s} \gamma_i^{-1} \right) \tag{3.34}$$

Since P_{gi}^0 $(i = N_s + 1, \ldots, N_g)$ have been chosen, equation (3.32) can be used to find P_D at every instant, as P_{di} are known at every instant from the system demand curve. Substituting P_D in (3.34) we get λ, and substituting λ in (3.33) we can arrive at the initial guess values P_{gi}^0 $(i = 2, \ldots, N_s)$ for the thermal stations. Note that P_{gi}^0 for the thermal buses will be different for different instants of time, as P_{di} vary with time, and hence P_D and λ.

Method for Updating P_{gi} for Successive Iterations. Our algorithm for generating new values P^n_{gi} starting from an initial value P^0_{gi} can be represented by the following successive approximation mapping:

$$P_{gi} = T(P_{gi}) \tag{3.35}$$

where T is a nonlinear operator representing the optimality equations. A natural way of generating a sequence of approximations is to use the new value of P_{gi} itself as the initial guess for the next iteration, i.e.,

$$P^{(k+1)}_{gi} = T[P^{(k)}_{gi}], \qquad k = 0, 1, 2, \ldots \tag{3.36}$$

This sequence may or may not converge depending on the form of T. If it does not converge we may try a modified mapping, by using

$$P^n_{gi} = 0.5[T(P^0_{gi}) + P^0_{gi}] \tag{3.37}$$

which leads to the sequence

$$P^{(k+1)}_{gi} = 0.5(TP^{(k)}_{gi} + P^{(k)}_{gi}), \qquad k = 0, 1, 2, \ldots \tag{3.38}$$

This means that if we start with the initial guess P^0_{gi} and arrive at a new set of values P^n_{gi}, the updated value for the next iteration would be $0.5(P^n_{gi} + P^0_{gi})$.

3.5. A Newton-Iterative Solution

An alternative solution technique for implementing HTOPF optimality conditions is via the use of Newton's iterative method with special features to deal with the large-scale nature of this complex problem.

This approach has been widely accepted in solving power systems problems owing to its reliable and fast convergence characteristics. Obtaining a solution usually takes only a few iterations provided that a resasonably good initial guess is used and that a solution exists.

In solving a multidimensional problem, Newton's method proceeds to find the $(k+1)$ vector of the unknowns on the basis of the prevous iteration's result x^k according to

$$X^{k+1} = X^k - \Delta X^k \tag{3.39a}$$

where

$$\Delta X^k = J^{-1}(X^k)F(X^k) \tag{3.39b}$$

The Jacobian matrix J is the matrix of partial derivatives of the functions f_i with respect to the unknown components x_i. Because of the large size of the problem treated, special attention should be paid to finding Δx^k using efficient linear equation solvers. Iterations proceed until there is no significant change in the correction vector ΔX^k.

In addition to the checking of ΔX^k, one can also check the mismatch vector $F(X^k)$. If the function values are small enough for the required accuracy, the iterative process can be terminated. Two different norms (i.e., L_2 and L_∞ norms) can be used as the stopping criterion. The L_2 norm is widely used in many areas. However, in the area of power systems, the L_∞ norm is conventionally used. The L_2 and L_∞ norms for the vector $X = (x_1, x_2, \ldots, x_n)^T$ are defined by

$$\|X\|_2 = \left(\sum_{i=1}^{n} x_i^2\right)^{1/2} \tag{3.40a}$$

and

$$\|X\|_\infty = \max_{1 \le i \le n} |x_i| \tag{3.40b}$$

3.5.1. The Jacobian

The optimality conditions for the hydrothermal optimal power flow problem are given by equation (3.11). The first six equations apply at each instant in time and only equation (3.11g) is time independent. To apply Newton's method to the problem it is helpful to classify the equations and the unknown variables as follows:

1a. Time-dependent functions:

$$f_s(t), f_h(t), f_\delta(t), f_v(t), f_p(t), f_q(t)$$

1b. Time-independent function:

$$f_\nu$$

2a. Time-dependent variables:

$$P_{gs}(t),\ P_{gh}(t),\ \delta(t),\ V(t),\ \lambda_P(t),\ \lambda_q(t)$$

2b. Time-independent variable:

$$\nu_i$$

Note that the following notations used for P_g are equivalent:

$$P_{gsi} \triangleq P_{gi}, \qquad i = 1, \ldots, N_s$$

$$P_{ghi} \triangleq P_{gi}, \qquad i = N_s + 1, \ldots, N_g$$

The Jacobian is formed by taking the partial derivatives of the functions with respect to the unknown variables. The element J_{xy} of the Jacobian is the partial derivative of the xth function with respect to the yth variable. In what follows we list the elements of the Jacobian.

The partial derivatives of f_s and f_h are obtained on the basis of equations (3.11a) and (3.11b):

$$\frac{\partial f_{si}}{\partial P_{gsi}} = 2\gamma_i, \qquad i = 1, \ldots, N_s \tag{3.41a}$$

$$\frac{\partial f_{hi}}{\partial P_{ghi}} = 2\gamma_i \nu_i, \qquad i = N_s + 1, \ldots, N_g \tag{3.41b}$$

$$\frac{\partial f_{si}}{\partial \lambda_{Pi}} = -1, \qquad i = 1, \ldots, N_s \tag{3.41c}$$

$$\frac{\partial f_{hi}}{\partial \lambda_{Pi}} = -1, \qquad i = N_s + 1, \ldots, N_g \tag{3.41d}$$

$$\frac{\partial f_{hi}}{\partial \nu_i} = \beta_i + 2\gamma_i P_{gi}, \qquad i = N_s + 1, \ldots, N_g \tag{3.41e}$$

The partial derivatives of f_δ and f_v are obtained on the basis of equations (3.11c) and (3.11d) as

$$\frac{\partial f_{\delta j}}{\partial \delta_k} = \sum_{i=1}^{N} \lambda_{Pi} \frac{\partial^2 P_i(V, \delta)}{\partial \delta_k \partial \delta_j} + \sum_{i=M+1}^{N} \lambda_{qi} \frac{\partial^2 Q_i(V, \delta)}{\partial \delta_k \partial \delta_j},$$

$$j = 2, \ldots, N, \qquad k = 2, \ldots, N \tag{3.42a}$$

$$\frac{\partial f_{\delta j}}{\partial V_k} = \sum_{i=1}^{N} \lambda_{Pi} \frac{\partial^2 P_i(V, \delta)}{\partial V_k \partial \delta_j} + \sum_{i=M+1}^{N} \lambda_{qi} \frac{\partial^2 Q_i(V, \delta)}{\partial V_k \partial \delta_j},$$

$$j = 2, \ldots, N, \qquad k = M + 1, \ldots, N \tag{3.42b}$$

$$\frac{\partial f_{vj}}{\partial \delta_k} = \sum_{i=1}^{N} \lambda_{Pi} \frac{\partial^2 P_i(V, \delta)}{\partial \delta_k \partial V_j} + \sum_{i=M+1}^{N} \lambda_{qi} \frac{\partial^2 Q_i(V, \delta)}{\partial \delta_k \partial \delta_j},$$

$$j = M + 1, \ldots, N, \qquad k = 2, \ldots, N \tag{3.42c}$$

$$\frac{\partial f_{vj}}{\partial V_k} = \sum_{i=1}^{N} \lambda_{Pi} \frac{\partial^2 P_i(V, \delta)}{\partial V_k \partial V_j} + \sum_{i=M+1}^{N} \lambda_{qi} \frac{\partial^2 Q_i(V, \delta)}{\partial V_k \partial V_j},$$

$$j = M + 1, \ldots, N, \qquad k = M + 1, \ldots, N \tag{3.42d}$$

It should be pointed out that the following relations between the Jacobian elements are true:

$$\frac{\partial f_{\delta j}}{\partial \delta_k} = \frac{\partial f_{\delta k}}{\partial \delta_j}, \qquad \frac{\partial f_{\delta j}}{\partial V_k} = \frac{\partial f_{Vk}}{\partial \delta_j}, \qquad \frac{\partial f_{vj}}{\partial V_k} = \frac{\partial f_{vk}}{\partial V_j}$$

Thus, the submatrix formed by these Jacobian elements is symmetric:

$$\frac{\partial f_{\delta j}}{\partial \lambda_{Pk}} = \frac{\partial P_k(V, \delta)}{\partial \delta_j}, \qquad j = 2, \ldots, N, \qquad k = 1, \ldots, N \tag{3.43a}$$

$$\frac{\partial f_{\delta j}}{\partial \lambda_{qk}} = \frac{\partial Q_k(V, \delta)}{\partial \delta_j}, \qquad j = 2, \ldots, N, \qquad k = M+1, \ldots, N \tag{3.43b}$$

$$\frac{\partial f_{vj}}{\partial \lambda_{Pk}} = \frac{\partial P_k(V, \delta)}{\partial V_j}, \qquad j = M+1, \ldots, N, \qquad k = 1, \ldots, N \tag{3.43c}$$

$$\frac{\partial f_{vj}}{\partial \lambda_{qk}} = \frac{\partial Q_k(V, \delta)}{\partial V_j}, \qquad j = M+1, \ldots, N, \qquad k = M+1, \ldots., N \tag{3.43d}$$

The submatrix formed by the above Jacobian elements is just the transpose of the load flow Jacobian of equations.

The partial derivatives of f_p and f_q are obtained on the basis of equations (3.11e) and (3.11f) as

$$\frac{\partial f_{Pi}}{\partial P_{gsi}} = -1, \qquad i = 1, \ldots, N_s \tag{3.44a}$$

$$\frac{\partial f_{Pi}}{\partial P_{ghi}} = -1, \qquad i = N_s + 1, \ldots, Ng \tag{3.44b}$$

$$\frac{\partial f_{Pi}}{\partial \delta_j} = \frac{\partial P_i(V, \delta)}{\partial \delta_j}, \qquad i = 1, \ldots, N, \qquad j = 2, \ldots., N \tag{3.44c}$$

$$\frac{\partial f_{Pi}}{\partial V_j} = \frac{\partial P_i(\delta, S)}{\partial V_j}, \qquad i = 1, \ldots, N, \qquad j = M+1, \ldots, N \tag{3.44d}$$

$$\frac{\partial f_{qi}}{\partial \delta_j} = \frac{\partial Q_i(V, \delta)}{\partial \delta_j}, \qquad i = M+1, \ldots, N, \qquad j = 2, \ldots, N \tag{3.44e}$$

$$\frac{\partial f_{qi}}{\partial V_j} = \frac{\partial Q_i(V, \delta)}{\partial V_j}, \qquad i = M+1, \ldots., N, \qquad j = M+1, \ldots, N \tag{3.44f}$$

The submatrix formed by the Jacobian elements of equations (3.44) is the Jacobian matrix of the load-flow equations.

The partial derivatives of f_ν are obtained on the basis of equation (3.11g) as

$$\frac{\partial f_{\nu i}}{\partial P_{ghi}} = (\beta_i + 2\gamma_i P_{gi})t_k, \qquad i = N_s + 1, \ldots, N_g, \qquad k = 1, \ldots, NT \quad (3.45)$$

The above expressions completely define all the Jacobian elements of the system of nonlinear equations derived from the optimality conditions of the augmented cost functional. The Jacobian matrix would be symmetric if the elements $f_{hi}/\partial \nu_i$ were equal to the elements $\partial f_{\nu i}/\partial P_{ghi}$. By examining the expressions for these elements, we see that they would be equal if and only if t_k had a value of unity for each time instant.

The structure of the Jacobian adopted in our implementation of Newton's method is shown in Fig. 3.1. This structure is given for only two time instants but the extension to the general case of N_T time instants is obvious.

3.5.2. Initial Guess Estimation

The convergence of Newton's iterations is sensitive to the initial guess of the unknown variables. We adopt a method that employs the given constraints to arrive at a physically based initial guess vector.

For the hydrogenerator it is assumed that longer time intervals or higher load demand will require more water. Thus, the prespecified volume of

	$P_{gs}(1)$	$P_{gh}(1)$	$P_{gs}(2)$	$P_{gh}(2)$	$\delta(1)$	$V(1)$	$\delta(2)$	$V(2)$	ν	$\lambda_p(1)$	$\lambda_q(1)$	$\lambda_p(2)$	$\lambda_q(2)$
$f_s(1)$	×	0	0	0	0	0	0	0	0	×	0	0	0
$f_h(1)$	0	×	0	0	0	0	0	0	×	×	0	0	0
$f_s(2)$	0	0	×	0	0	0	0	0	0	0	0	×	0
$f_h(2)$	0	0	0	×	0	0	0	0	×	0	0	×	0
$f_\delta(1)$	0	0	0	0	×	×	0	0	0	×	×	0	0
$f_v(1)$	0	0	0	0	×	×	0	0	0	×	×	0	0
$f_\delta(2)$	0	0	0	0	0	0	×	×	0	0	0	×	×
$f_v(2)$	0	0	0	0	0	0	×	×	0	0	0	×	×
f_ν	0	×	0	×	0	0	0	0	0	0	0	0	0
$f_p(1)$	×	×	0	0	×	×	0	0	0	0	0	0	0
$f_q(1)$	0	0	0	0	×	×	0	0	0	0	0	0	0
$f_p(2)$	0	0	×	×	0	0	×	×	0	0	0	0	0
$f_q(2)$	0	0	0	0	0	0	×	×	0	0	0	0	0

Figure 3.1. Jacobian structure for the HTOPF.

water over the optimization period considered can be allotted into each time interval based on the above two factors (i.e., interval duration and load demand level):

$$b_{ik} = b_i \frac{C_k T_k}{CT}, \qquad i \in R_H, \qquad k = 1, \ldots, NT \tag{3.46}$$

$$CT = \sum_{k=1}^{NT} C_k T_k \tag{3.47}$$

where b_{ik} is the volume of water used by the kth time interval.

Since the hydrogeneration is assumed constant at every time interval, the discharge q_{ik} is also constant. Thus, the volume of water is represented by

$$q_{ik} T_k = b_{ik}, \qquad i \in R_H, \qquad k = 1, \ldots, NT \tag{3.48}$$

Substituting equation (3.46) into (3.47), and equation (3.10b) for q_{ik}, we get

$$(\alpha_i + \beta_i P_{gik} + \nu_i P_{gik}^2) T_k = b_i \frac{C_k T_k}{CT}, \qquad i \in R_H, \qquad k = 1, \ldots, NT$$

As a result we obtain

$$\alpha_i + \beta_i P_{gik} + \gamma_i P_{gik}^2 = b_i \frac{C_k}{CT}, \qquad i \in R_H, \qquad k = 1, \ldots, NT$$

A quadratic formula is used to solve for P_g's and the positive root is chosen:

$$P_{gik} = \frac{-\beta_i + [\beta_i^2 - 4\gamma_i(\alpha_i - q_{ik})]^{1/2}}{2\gamma_i}, \qquad i \in R_H, \qquad k = 1, \ldots, NT \tag{3.49}$$

We can now obtain the remaining initial guess values

$$P_D - \sum_{i \in R_s} P_{gi} = \sum_{i=1}^{N} P_{di} - \sum_{i \in R_H} P_{gi} \tag{3.50}$$

$$P_{gi} = \frac{1}{2\gamma_i}(\lambda - \beta_i), \qquad i \in R_s \tag{3.51}$$

$$\lambda = \frac{2P_D + \sum_{i \in R_s} (\beta_i / \gamma_i)}{\sum_{i \in R_s} (1/\gamma_i)} \tag{3.52}$$

$$\lambda_{Pi} = \lambda, \qquad i = 1, \ldots, N \tag{3.53a}$$

$$\lambda_{qi} = 0, \qquad i = M+1, \ldots, N \tag{3.53b}$$

$$\delta_i = 0, \qquad i = 2, \ldots, N \tag{3.54a}$$

$$V_i = 1, \qquad i = M+1, \ldots, N \tag{3.54b}$$

$$\nu_i = \frac{\lambda_{pi}}{\beta_i + 2\gamma_i P_{gi}}, \qquad i \in R_H \tag{3.55}$$

Note that this method generates the initial values for all the variables required by Newton's algorithm.

3.6. Computational Performance

The iterative techniques of Sections 3.4 and 3.5 were tested for standard test systems varying in size from a five-bus system to a 118-bus system. Presently, we summarize test results.

3.6.1. Successive Approximation Algorithm

This algorithm was tested for three systems (Ref. 3.18), of size up to 30 buses.

Test System 1. This is a standard five-bus system taken from Stagg and El-Abiad (Ref. 3.1). Two generators are involved. Generator 1 is thermal, and Generator 2 is a hydrogenerator. The fuel cost of the thermal generator is assumed to be

$$F_1 = 52.022 + 10.515P_1 + 0.011138P_1^2 \text{ \$/hr}$$

and the rate of discharge at the hydrogenerator is assumed to be

$$q_2 = 8.5234 + 1.6982P_2 + 0.18277 \times 10^{-2}P_2^2 \text{ m}^3/\text{s}$$

P_1 and P_2 are in MW. The volume of water available is taken as $b = 14.158$ million cubic meters.

The line parameters are given in Table 3.1. The daily load curve has a peak demand of 1 p.u. lasting for eight hours. Demand is 0.7 p.u. for the first eight hours and 0.9 p.u. for the last eight hours. There are time segments of eight hours width over which the generation and demand are

Table 3.1. Five-Bus Test System of Stagg and El-Abiad, Impedance and Line Charging Data[a]

Line designation		Resistance	Reactance	Line charging
1	2	0.02	0.06	0.030
1	3	0.08	0.24	0.025
2	3	0.06	0.18	0.020
2	4	0.06	0.12	0.015
2	5	0.04	0.12	0.010
3	4	0.01	0.03	0.025
4	5	0.08	0.24	0.025

[a] Impedances and line charging susceptances are per unit on a 100-MVA base. Line charging one-half of the total charging of line.

assumed constant. The actual demands at the various buses at peak time are given in Table 3.2. The generator bus voltages $|V_1|$ and $|V_2|$ are assumed to be constant at all times: $|V_1| = 1.06000$ p.u. and $|V_2| = 1.04748$ p.u. Bus 1 is taken as the slack bus. The system is treated as unconstrained as far as the reactive power generation is concerned. The initial guess value for P_{g2} is chosen as 83.9 MW over all the three time segments; the value was obtained using equation (3.31). The program was run on the University of Alberta computer, an Amdahl 470V/6. P_{g2} converged to the values 68.562, 95.992, and 86.793 over the three segments of time 0–8 hr, 8–16 hr, and 16–24 hr, respectively, to an accuracy of 0.001 in 22 iterations. The equivalent CPU time was 1.759 s. Convergence to an accuracy of 0.01 required only seven iterations corresponding to an equivalent CPU time of 0.769 s. The results of the run, with convergence accuracy set at 0.001, are given in Table 3.3. The table shows the optimal active and reactive power generation at the thermal and hydrobuses, as well as the voltage magnitudes and angles at all buses, over the three time segments considered. The data are sufficient to compute the optimal line flows if required.

Test System 2. This is a 14-bus system designated as the IEEE-AEP System. As before there are two generators:

Generator 1 (Thermal):

$$F_1 = 53.607 + 10.662P_1 + 0.01165P_1^2 \ \$/\text{hr}$$

Generator 2 (Hydro):

$$q_2 = 2.8317 + 0.156610P_2 + 0.060923 \times 10^{-2}P_2^2 \ \text{m}^3/\text{s}$$

P_1 and P_2 are in MW. It is assumed that $b = 2.8317$ million cubic meters over 24 hr.

System data are given in Table 3.4. The daily load curve has a peak of 1 p.u. for four hours and is 0.5 p.u. for the rest of the time. The actual demands at peak time at the various buses are give in Table 3.5. Though there are three steps in the load curve, we need only consider two segments of time of width 4 and 20 hr corresponding to the time intervals 8–12 hr

Table 3.2. Five-Bus Test System of Stagg and El-Abiad, Peak Demands at the Buses

Bus No.	P_d (MW)	Q_d (MVAR)
1	0	0
2	20	10
3	45	15
4	40	5
5	60	10

Table 3.3. Five-Bus Test System of Stagg and El-Abiad, Optimal Solution

Time segment (hr)	Bus No.	Voltage magnitude $\lvert V \rvert$ (V)	Angle δ (rad)	Optimal P_g (MW)	Generation Q_g (MVAR)
0–8	1	1.06000	0.0000	48.691	11.322
	2	1.04748	−0.01205	68.562	−7.163
	3	1.03504	−0.04496		
	4	1.03455	−0.04783		
	5	1.02948	−0.05449		
8–16	1	1.06000	0.00000	72.640	8.827
	2	1.04748	−0.02032	95.992	9.779
	3	1.02449	−0.06602		
	4	1.02380	−0.07030		
	5	1.01793	−0.08064		
16–24	1	1.06000	0.00000	64.518	9.604
	2	1.04748	−0.01756	86.793	3.955
	3	1.02806	−0.05896		
	4	1.02741	−0.06277		
	5	1.02188	−0.07186		

and 0–8, 12–24 hr, over which demand the generations are constant at full peak and one-half peak values, respectively. Note that there are five regulated (voltage-controlled) buses at which voltages are sought to be held constant, as long as the inequality constraints on reactive power generation are not violated. These are the generator buses 1 and 2, and the synchronous condenser buses 3, 6, and 8. For the purpose of computation, the synchronous condenser buses are treated as generator buses with zero active power generation and positive reactive power generation. Also, the effect of the static capacitor in bus 9 and the 3 tap-changing transformers is included in the nodal admittance matrix by suitably modifying it. As before, bus 1 is taken as the slack bus. The initial guess value for P_{g2} is computed as 50 MW from equation (3.31) and the program run on the computer. P_{g2} converged to the values 104.71 and 38.51 over the segments 8–12 hr and 0–8, 12–24 hr to a convergence accuracy of 0.01 in six iterations. The equivalent CPU time was 13.738 s. The results of the run are given in Table 3.6. It may be noted that the inequality constraints were violated at buses 6 and 8 over the time segment 8–12 h and at buses 3 and 6 over the time segment 0–12, 12–24 hr; consequently the reactive power generation at these buses was set to the violated limits over the respective time segments, and the voltages are no longer the specified values.

Test System 3. This is a 30-bus system known as the IEEE-AEP 30 bus system. There are 41 lines, two thermal generators, and one hydroplant.

Table 3.4. 14-Bus IEEE-AEP System[a]

(a) Impedance and Line Charging Data

Line	Designation	Resistance	Reactance	Line charging
1	2	0.01938	0.05917	0.02640
2	3	0.04699	0.19797	0.02190
2	4	0.05811	0.17632	0.01870
1	5	0.05403	0.22304	0.02460
2	5	0.05695	0.17388	0.01700
3	4	0.06701	0.17103	0.01730
4	5	0.01335	0.04211	0.00640
5	6	0.00000	0.25202	0.00000
4	7	0.00000	0.20912	0.00000
7	8	0.00000	0.17615	0.00000
4	9	0.00000	0.55618	0.00000
7	9	0.00000	0.11001	0.00000
9	10	0.03181	0.08450	0.00000
6	11	0.09498	0.19890	0.00000
6	12	0.12291	0.25581	0.00000
6	13	0.06615	0.13027	0.00000
9	14	0.19711	0.27038	0.00000
10	11	0.08205	0.19207	0.00000
12	13	0.22092	0.19988	0.00000
13	14	0.17093	0.34802	0.00000

(b) Regulated Bus Data

Bus No.	Voltage magnitude per unit	Minimum MVAR capability	Maximimum MVAR capability
1	1.060	—	—
2	1.045	−40.0	50.0
3	1.010	0.0	40.0
4	1.070	−6.0	24.0
5	1.090	−6.0	24.0

(c) Transformer Data

Transformer	Designation	Tap setting
4	7	0.978
4	9	0.969
5	6	0.932

(d) Shunt Data

Shunt designation	Susceptance per unit
9	0.19000

[a] Impedances and line charging susceptances are per unit on a 100-MVA base. Line charging one-half of the total charging of line.

Table 3.5. 14-Bus IEEE-AEP System, Peak Demands at the Buses

Bus No.	P_d (MW)	Q_d (MVAR)	Bus No.	P_d(MW)	Q_d (MVAR)
1	0.0	0.0	8	0.0	0.0
2	21.7	12.7	9	29.5	16.6
3	94.2	19.0	10	9.0	5.8
4	47.8	−3.9	11	3.5	1.8
5	7.6	1.6	12	6.1	1.6
6	11.2	7.5	13	13.5	5.8
7	0.0	0.0	14	14.9	5.0

Table 3.6. 14-Bus IEEE-AEP System, Optimal Solution

Time segment (hr)	Bus No.	Voltage magnitude $\|V\|$ (V)	Angle δ (rad)	Optimal P_g (MW)	Generation Q_g (MVAR)
8–12	1	1.060	0.000	162.11	−8.34
	2	1.045	−0.053	104.71	1.33
	3	1.010	−0.190		16.12
	4	1.031	−0.155		
	5	1.036	−0.131		
	6	0.995	−0.221		24.00
	7	1.035	−0.211		
	8	1.074	−0.211		24.00
	9	1.016	−0.241		
	10	1.005	−0.243		
	11	0.996	−0.235		
	12	0.982	−0.238		
	13	0.981	−0.240		
	14	0.987	−0.258		
0–8, 12–24	1	1.060	0.000	97.10	−3.68
	2	1.045	−0.031	38.51	−35.92
	3	1.032	−0.099		0.00
	4	1.053	−0.086		
	5	1.057	−0.074		
	6	1.026	−0.119		24.00
	7	1.054	−0.116		
	8	1.090	−0.116		22.15
	9	1.046	−0.133		
	10	1.039	−0.133		
	11	1.031	−0.128		
	12	1.019	−0.128		
	13	1.017	−0.129		
	14	1.013	−0.138		

The network includes two synchronous condensers and two static capacitors as well as four on-line tap changers. Buses 1 and 13 are thermal-generation buses and bus 2 is a hydrogeneration bus.

System data are given in Table 3.7 and the peak demands at the system load buses are shown in Table 3.8. The load curve for the system is the same as that for test system 2. The initial guess on P_{g1} for the first interval is 98 MW and 101 MW for the second interval. The optimal solution was obtained to an accuracy of 0.01 in an equivalent CPU time of 3 min and 65.4 s (see Table 3.9).

Remarks:

1. Memory requirements for this algorithm are a modest 10K but obviously data storage requirements are strongly dependent on the number of buses in the system and the number of discrete time intervals. For the 5-bus system, about 5K of storage is required, whereas the 30-bus system requires close to 95K of storage. As the system size increases, data storage is increased.
2. Computational time increases sixfold for a doubling of system size. This is due to two factors:
   a. Additional time required for swapping data in and out of active memory.
   b. Increase in incidence of inequality constraint violations.
3. For a system with slightly over 100 buses, computational time for two iterations took about two CPU hours. This indicated that a realistic limit for this algorithm is close to 37 buses from a computational time point of view.

3.6.2. Newton's Method

Newton's method for the HTOPF problem was implemented in a computer program written in VAX-11 FORTRAN, which is based on the American National Standard FORTRAN-77, and tested on a VAX-750 using double precision arithmetic. The solution of a sparse system of linear equations required for updating the solution was performed using Harwell's Library routine MA28A, which uses a variant of Gaussian elimination for sparse systems.

The program was employed to solve five test systems varying in size from five buses to 57 buses. Table 3.10 lists the characteristics of these test systems. The load curve for the system tested was assumed to be 0.5 per unit for 20 hr and one per unit for 4 hr. In the case of the five-bus system, three discrete intervals with 0.7 p.u. for 8 hr were taken for the load curve. The actual load in MW and MVAR at each load bus is found by appropriate scaling of the peak values given in the test system specifications.

Table 3.7. 30-Bus IEEE-AEP System

(a) Impedance, Line Charging, and Transformer Tap Setting

Line designation			Resistance	Reactance	Line charging	Tap setting
1	1	2	0.01920	0.05750	0.02640	
2	1	3	0.04520	0.13520	0.02040	
3	2	4	0.05700	0.17370	0.01840	
4	3	4	0.01320	0.03790	0.00420	
5	2	5	0.04720	0.19830	0.02090	
6	2	6	0.05810	0.17630	0.01879	
7	4	6	0.01190	0.04140	0.00450	
8	5	7	0.04600	0.11600	0.01020	
9	6	7	0.02670	0.08200	0.00850	
10	6	8	0.01200	0.04200	0.00450	
11	6	9	0.00000	0.20800	0.00000	0.97800
12	6	10	0.00000	0.55600	0.00000	0.96900
13	9	11	0.00000	0.20800	0.00000	
14	9	10	0.00000	0.11000	0.00000	
15	4	12	0.00000	0.25600	0.00000	0.93200
16	12	13	0.00000	0.14000	0.00000	
17	12	14	0.12310	0.25590	0.00000	
18	12	15	0.06620	0.13040	0.00000	
19	12	16	0.09450	0.19870	0.00000	
20	14	15	0.22100	0.19970	0.00000	
21	16	17	0.08240	0.19230	0.00000	
22	15	18	0.10730	0.21850	0.00000	
23	18	19	0.06390	0.12920	0.00000	
24	19	20	0.03400	0.06800	0.00000	
25	10	20	0.09360	0.20900	0.00000	
26	10	17	0.03240	0.08450	0.00000	
27	10	21	0.03480	0.07490	0.00000	
28	10	22	0.07280	0.14990	0.00000	
29	21	22	0.01160	0.02360	0.00000	
30	15	23	0.10000	0.20200	0.00000	
31	22	24	0.11500	0.17900	0.00000	
32	23	24	0.13200	0.27000	0.00000	
33	24	25	0.18850	0.32920	0.00000	
34	25	26	0.25440	0.38000	0.00000	
35	25	27	0.10930	0.20870	0.00000	
36	28	27	0.00000	0.39600	0.00000	0.96800
37	27	29	0.21980	0.41530	0.00000	
38	27	30	0.32020	0.60270	0.00000	
39	29	30	0.23990	0.45330	0.00000	
40	8	28	0.06360	0.20000	0.02140	
41	6	28	0.01690	0.05990	0.00650	

continued overleaf

Table 3.7 (*continued*)

(*b*) *Regulated Bus Data*

Bus number	Voltage magnitude per unit	Minimum MVAR capability	Maximum MVAR capability
1	1.06000	0.00000	0.00000
2	1.04300	−40.00000	50.00000
5	1.01000	−40.00000	40.00000
11	1.08200	−6.00000	30.00000
13	1.07100	−10.00000	50.00000

(*c*) *Shunt Data*

Shunt designation	Susceptance per unit
10	0.190
24	0.043

The thermal cost models used are given by

$$F = 52.022 + 10.5125P_g + 0.011138P_g^2 \qquad \text{(5-bus system)}$$

$$F = 53.607 + 10.662P_g + 0.01165P_g^2 \qquad \text{(all other systems)}$$

The hydroplant performance models are given by

$$q = 1.0836 + 0.2159P_g + 2.3236 \times 10^{-4}P_g^2 \qquad \text{(5-bus system)}$$

$$P = 0.36 + 0.07197P_g + 0.77435 \times 10^{-4}P_g^2 \qquad \text{(all other systems)}$$

The algorithm performed well in obtaining the optimal solution for the systems tested. A summary of the results is given below.

5-Bus System

Convergence to the optimal solution in five iterations was obtained. Bus voltage magnitudes ranged from 1.06 at bus 1 to 1.018. The optimal, ν, is found as 59.847. Table 3.11 lists the optimal active and reactive power generations and Table 3.12 lists the error measures, ε_1 and ε_2, at each itertion. A program area of 13.28K was required and data areas of 24.26K for three time intervals and 1.13M for 24 time intervals were used.

14-Bus System

Convergence in five iterations to a solution that violated two constraints was obtained. Subsequent voltage-controlled bus release and restart resulted in convergence in five iterations. Voltage magnitudes in the range 1.06–0.956 were obtained. Optimal, ν, is found as 105.96. Table 3.13 lists the optimal

Table 3.8. 30-Bus IEEE-AEP System, Peak Demand at the Buses

Bus No.	P_d (MW)	Q_d (MVAR)
1	0.0	0.0
2	21.7	12.7
3	2.4	1.2
4	7.6	1.6
5	94.2	19.0
6	0.0	0.0
7	22.3	10.9
8	30.0	30.0
9	0.0	0.0
10	5.8	2.0
11	0.0	0.0
12	11.2	7.5
13	0.0	0.0
14	6.2	1.6
15	8.2	2.5
16	3.5	1.8
17	9.0	5.8
18	3.2	0.9
19	9.5	3.4
20	2.2	0.7
21	17.5	11.2
22	0.0	0.0
23	3.2	1.6
24	8.7	6.7
25	0.0	0.0
26	3.5	2.3
27	0.0	0.0
28	0.0	0.0
29	2.4	0.9
30	10.5	1.9

active and reactive power generation and Table 3.14 lists the error measures, ε_1 and ε_2, at each iteration. The data area for two intervals is 82.76K and 9.29M for 24 hr.

30-Bus System

Convergence was obtained in five iterations with a subsequent voltage-controlled bus release at one bus. Four more iterations produced the optimal solution shown in Table 3.15. Voltage magnitudes of 1.08–0.993 were obtained. Optimal, ν, is 152.582. Table 3.16 lists the error measures for each iteration. A data area for 24 hr of 44M was required.

Table 3.9. 30-Bus IEEE-AEP System, Optimal Solution

Bus No.	Voltage magnitude $\lvert V \rvert$ (V)	Angle δ (rad)	Optimal P_g (MW)	Generation Q_g (MVAR)
		First Interval		
1	1.0600000	0.0000000	97.58	93.67
2	1.0430000	−0.0304751	93.63	−1.85
3	1.0410929	−0.0577634		
4	1.0361029	−0.0687951		
5	1.0100000	−0.1645265		31.97
6	1.0202524	−0.0933796		
7	1.0083509	−0.1316340		
8	1.0067323	−0.1020642		
9	1.0143113	−0.1123238		
10	0.9928296	−0.1236283		
11	1.0724984	−0.1128237		30.00
12	1.0149086	−0.0197965		
13	1.0710000	−0.1118438	101.91	49.63
14	0.9989364	−0.0479713		
15	0.9894736	−0.0608446		
16	0.9954963	−0.0682254		
17	0.9872809	−0.1108696		
18	0.9770281	−0.0962154		
19	0.9733964	−0.1137394		
20	0.9772207	−0.1173964		
21	0.9792889	−0.1301117		
22	0.9797123	−0.1291567		
23	0.9762807	−0.0916725		
24	0.9665393	−0.1269227		
25	0.9560402	−0.1431607		
26	0.9371804	−0.1514742		
27	0.9592287	−0.1483178		
28	1.0140680	−0.1004189		
29	0.9379043	−0.1728148		
30	0.9255778	−0.1904791		
		Second Interval		
1	1.0600000	0.0000000	50.83	12.79
2	1.0430000	−0.0137089	41.04	−30.65
3	1.0571316	−0.0301364		
4	1.0553846	−0.0357711		
5	1.0100000	−0.0735403		
6	1.0461705	−0.0476077		
7	1.0281879	−0.0631264		
8	1.0404732	−0.0513741		
9	1.0449134	−0.0569040		
10	1.0378687	−0.0619738		
11	1.0820000	−0.0569039		19.29
12	1.0964571	−0.0082678		
13	1.0710000	−0.0580203	52.52	28.167
14	1.0305499	−0.0221497		

Table 3.9 (*continued*)

Bus No.	Voltage magnitude $\|V\|$ (V)	Angle δ (rad)	Optimal P_g (MW)	Generation Q_g (MVAR)
	Second Interval (continued)			
15	1.0274309	− 0.0293308		
16	1.0322323	− 0.0335316		
17	1.0328239	− 0.0552980		
18	1.0245749	− 0.0469999		
19	1.0245535	− 0.0557918		
20	1.0272493	− 0.0578976		
21	1.0305896	− 0.0649370		
22	1.0306789	− 0.0645083		
23	1.0235191	− 0.0452098		
24	1.0229264	− 0.0633832		
25	1.0112864	− 0.0692324		
26	1.0024794	− 0.0729060		
27	1.0085254	− 0.0708299		
28	1.0451358	− 0.0511558		
29	0.9987180	− 0.0817730		
30	0.9930341	− 0.0895078		

Table 3.10. Characteristics of IEEE Standard Test Systems

Number of nodes	5	14	30	37	57
Number of lines	7	20	41	46	78
Number of thermal generators	1	1	2	2	2
Number of hydrogenerators	1	1	1	2	2
Number of synchronous condensers	0	3	2	0	3
Number of static capacitors	0	1	2	0	3
Number of transformers	0	3	4	0	15

Table 3.11. Optimal Solution for Five-Bus System

Generator	$P_G(1)$	$Q_G(1)$	$P_G(2)$	$Q_G(2)$	$P_G(3)$	$Q_G(3)$
1	48.516	11.385	72.28	8.88	64.34	9.664
2	68.560	−10.183	95.99	9.75	86.79	2.927

Table 3.12. Variation of Error Measures with Iterations for Five-Bus System

Measure	Iteration: 1	2	3	4	5
ε_1	68.62	9.81	0.027	0.5×10^{-6}	0.26×10^{-12}
ε_2	156.60	29.10	0.099	0.13×10^{-5}	0.54×10^{-12}

Table 3.13. Optimal Solution for 14-Bus System

Bus	$P_G(1)$	$Q_G(1)$	$P_G(2)$	$Q_G(2)$
1	162.433	−5.293	94.67	−4.050
2	107.233	9.186	37.99	−38.990
3		20.63		0.000
6		24.00		24.000
8		24.00		20.216

Table 3.14. Variation of Error Measures with Iterations for 14-Bus System

Measure	Iteration: 1	2	3	4	5
ε_1	122.62	381.50	0.41	0.766×10^{-4}	0.139×10^{-11}
ε_2	348.96	152.88	1.61	0.217×10^{-3}	0.440×10^{-11}

Table 3.15. Optimal Solution for 30-Bus System

Bus	$Q_G(1)$	$Q_G(1)$	$P_G(2)$	$Q_G(2)$
1	97.580	9.367	50.88	12.790
13	101.911	49.627	52.52	28.167
2	93.626	−1.849	41.046	−30.640
11		30.000		19.292
5		31.970		−11.972

Table 3.16. Variation of Error Measures with Iterations for 30-Bus System

Measure	Iteration: 1	2	3	4	5
ε_1	146.94	38.96	1.540	0.147×10^{-2}	0.61×10^{-9}
ε_2	486.82	123.24	3.705	0.275×10^{-2}	0.101×10^{-8}

Table 3.17. Optimal Solution for 37-Bus System

Bus	$P_G(1)$	$Q_G(1)$	$P_G(2)$	$Q_G(2)$
1	841.422	57.06	452.154	−137.634
8	552.536	50.00	391.537	50.000
7	164.115	60.00	24.212	59.900
2	218.010	100.00	9.540	100.000

37-Bus System

Convergence in six iterations with subsequent voltage control release at three buses resulted in optimal solution in five iterations. Voltage magnitudes in the range 1.13–0.785 were obtained. Optimal, ν, at bus 7 is 278.77 and that of bus 2 is 292.113. Table 3.17 shows the optimal active and reactive generations and Table 3.18 shows the error measure variation with iterations. For two intervals a data area of 555K was required. For 24 hr we require a data area of 67.8M.

Table 3.18. Variation of Error Measure with Iteration Number

Measure	Iteration: 1	2	3	4	5	6
ε_1	720.84	699.61	119.18	1.23	0.586×10^{-4}	0.37×10^{-11}
ε_2	1369.22	1119.6	144.79	1.45	0.697×10^{-4}	0.89×10^{-11}

Table 3.19. Optimal Solution for 57-Bus System

Bus	$P_G(1)$	$Q_G(1)$	$P_G(2)$	$Q_G(2)$
1	266.760	39.78	148.825	34.47
3	809.686	−10.00	469.672	−10.00
8	148.360	200.00	28.205	44.99
12	145.038	155.00	29.029	76.07
2		50.00		50.00
9		9.00		−3.00
6		25.00		−8.00

Table 3.20. Variation of Error Measures with Iteration Count for 57-Bus System

Measure	Iteration: 1	2	3	4	5	6
ε_1	792.12	826.95	236.98	10.37	0.168×10^{-1}	0.21×10^{-7}
ε_2	5452.90	5251.53	909.77	29.07	0.402×10^{-1}	0.5×10^{-7}

Table 3.21. CPU Time Requirements (hr:min:s) on VAX-750 for Systems Tested

	No. of buses: 5	14	30	37	57
(1)	5.91	38.75	3:38.79	6:36.66	41:23.25
(2)	6.59	46.45	3:57.66	7:06.64	45:27.23

57-Bus System

An initial attempt with six iterations resulted in a solution with violations at five buses in interval one and three buses in interval two. Fix up and six more iterations still resulted in violations at six buses in interval one and four buses in interval two. Seven more iterations with voltage-control bus release resulted in the obtained solution listed in Table 3.19. Optimal, ν, for bus 8 is 323.77 and for bus 12 is 325.237. Optimal voltage magnitudes in the range 1.04–0.844 were obtained. A data area of 15M for 24 hr is required, whereas a data area of 1.29M is required for two intervals. Table 3.20 lists the variation of the error measures for the iterations.

Computational Time

Table 3.21 lists the CPU time required on the VAX-750 to solve the test systems. As expected, inordinate computational time is required for implementing the procedure for large systems.

3.7. Conclusions

A major optimal control problem in electric power systems operations has been treated in this chapter. Work reported here encompasses the formulation of the optimal load flow problem in a hydrothermal electric power system. The variational solution to this dynamic optimization problem results in a large-scale system of nonlinear equations which must be solved using iterative methods.

The successive approximation method requires less memory, but can be slow to converge as compared to Newton's method. The latter holds promise for large systems.

It is important to realize that an increase in system size results in an increased storage and computational time burden. Although it can be argued that studies of large systems involving dynamics (such as this one) can only be practical for reduced equivalents of reasonable size, our results show the limitations of the present algorithm. As a result, the following areas of improvement should be actively researched:

1. Improve the initial guess generation procedure.
2. Seek alternative forms of iterative formulas that improve convergence characteristics.
3. Seek to exploit the relationships between variables over successive time intervals to reduce the dependence on the number of time intervals.

References

3.1. STAGG, G. W., and EL-ABIAD, A. H., *Computer Methods in Power System Analysis*, McGraw-Hill, New York, 1968.

3.2. EL-HAWARY, M. E., *Electrical Power Systems: Design and Analysis*, Prentice-Hall (A Reston Book), Englewood Cliffs, New Jersey, 1983.

3.3. BERGEN, A. R., *Power Systems Analysis*, Prentice-Hall, Englewood Cliffs, New Jersey, 1986.

3.4. KUSIC, G. L., *Computer-Aided Power Systems Analysis*, Prentice-Hall, Englewood Cliffs, New Jersey, 1986.

3.5. SQUIRES, R. B., "Economic Dispatch of Generation Directly from Power System Voltages and Admittances," *AIEE Trans. Power Appar. Syst.* **PAS-79**(III), 1235–1244 (1961).

3.6. CARPENTIER, J., "Contribution a l'étude du Dispatching Economique," *Bull. Soc. Fr. Elec. Ser. B* **3**, 431–447 (1962).

3.7. SUN, D. I., ASHLEY, B., BREWER, B., HUGHES, A., and TINNEY, W. F., "Optimal Power Flow by Newton Approach," *IEEE Trans. Power Appar. Syst.* **103**(10), 2864–2880 (1984).

3.8. BURCHETT, R. C., HAPP, H. H., PALMER, R. E., and VIERATH, D. R., "Quadratically Convergent Optimal Power Flow," *IEEE Trans. Power Appar. Syst.* **103**(11), 3267–3275 (1984).

3.9. EL-KADY, M. A., BELL, B. D., CARVALHO, V. F., BURCHETT, R. C., HAPP, H. H., and VIERATH, D. R., "Assessment of Real Time Optimal Voltage Control," Paper 85-SM-489-0, 1985 Summer Power Meeting, Vancouver.

3.10. RICARD, J., "The Determination of Optimum Operating Schedule for Inter-Connected Hydro and Thermal Stations," *Rev. Gen. Electr. Paris*, 167 (1940).

3.11. CHANDLER, W. G., DANDENO, P. L., GLIMM, A. F., and KIRCHMAYER, L. K., "Short Range Economic Operation of a Combined Thermal and Hydro-Electric Power System," *AIEE Trans.* **72**(III), 1057–1065 (1953).

3.12. GLIMM, A. F., and KIRCHMAYER, L. K., "Economic Operation of Variable-Head Hydro Electric Plants," *AIEE Trans.* **77**(III), 1070–1079 (1958).

3.13. RAMAMOORTHY, M., and GOPALA RAO, J., "Load Scheduling of Hydro-Electric Thermal Generating Systems Using Nonlinear Programming Techniques," *Proc. IEEE* **117**, 794–798 (1970).

3.14. EL-HAWARY, M. E., and CHRISTENSEN, G. S., "Hydro-Thermal Load Flow Using Functional Analysis," *J. Optimization Theory Appl.* **12**, 576–587 (1973).

3.15. EL-HAWARY, M. E., and CHRISTENSEN, G. S., *Optimal Economic Operation of Electric Power Systems*, Academic, New York, 1979.

3.16. KIRCHMAYER, L. K., *Economic Operation of Power Systems*, Wiley, New York, 1958.

3.17. WOOD, A. J., and WOLLENBERG, B., *Power Generation, Operation, and Control*, Wiley, New York, 1984.

3.18. EL-HAWARY, M. E., RAO, R. S., and CHRISTENSEN, G. S., "Optimal Hydro-Thermal Load Flow: Formulation and a Successive Approximation Solution for Fixed Head Systems," *J. Optimal Control Appl. Meth.* **7**(4), 337–355 (1986).

3.19. EL-HAWARY, M. E., and TSANG, D. H., "The Hydro-Thermal Optimal Load Flow: A Practical Formulation and Solution Technique Using Newton's Approach," *IEEE Trans. Power Syst. Eng.* **PWRS-1**(3), 157–167 (1986).

4

Economic Coordination of Hydrothermal–Nuclear Systems

4.1. Introduction

It is the purpose of this chapter (1) to formulate the problem of optimal short-term operation of hydrothermal-nuclear systems, (2) to obtain the solution by using a functional analytic optimization technique that employs the minimum norm formulation, and (3) to propose an algorithm suitable for implementing the optimal solution.

At present, owing to economic considerations, most of the nuclear generating stations throughout the world are operated at base load, and the length of the reactor refueling cycles for most reactors is of the order of several years. Thus it is recognized that the load scheduling of mixed power systems with nuclear plants is essentially an optimization problem of long-term nature (Refs. 4.1–4.3).

The multiyear length of the optimization interval makes it virtually impossible to consider the hourly or even daily load variations of the power demand.

With nuclear plants becoming an increasing part of power generating systems it is expected that in the near future the operating role of the nuclear plants will change. It is likely that in the absence of suitable energy storage methods the nuclear plants will be required to operate in a load-following mode in order to achieve the flexibility needed to match the power demand (Ref. 4.4).

The scope of the present chapter is limited to cover power systems with thermal nuclear reactors (Ref. 4.5) that are provided with continuous on power refueling capabilities (Ref. 4.6).

In this case the long-term effect of reactor fueling cycles is not present (Ref. 4.7) and the short-term scheduling becomes a meaningful problem.

The functional analytic optimization technique that is used here has been applied successfully to the analysis of hydrothermal systems (Ref. 4.8). Essentially this technique transforms the optimization problem into a minimum norm problem subject to a linear transformation in the context of functional analysis (Ref. 4.9). One of the advantages of this method is that it eliminates the multipliers associated with linear constraints.

The organization of the material in this chapter is as follows: in Section 4.2 the problem is defined and the development of the system's model is presented; in Section 4.3 the optimization problem is reformulated in the context of functional analysis; Section 4.4 presents the solution to the problem; and Section 4.5 proposes an iterative method for its implementation.

4.2. Problem Formulation

The power system considered contains N_N nuclear plants, N_T fossil fueled plants, and N_H hydrogenerating stations located on the same stream. The problem is to determine the load schedule such that the total operating costs of the system in a time interval T are minimized under the following conditions:

1. All operating costs are attributed to the nuclear and fossil fuels. The total operating cost is assumed of the form

$$C = \int_0^T \left[\sum_{i=1}^{N_N} (\alpha_{Ni} + \beta_{Ni} P_{Ni} + \gamma_{Ni} P_{Ni}^2) + \sum_{i=1}^{N_T} (\alpha_{Ti} + \beta_{Ti} P_{Ti} + \gamma_{Ti} P_{Ti}^2) \right] dt \tag{4.1}$$

where P_{Ni} and P_{Ti} are the electric power generations at the ith nuclear and thermal plants, respectively. The coefficients α_i, β_i, and γ_i are known constants.

2. The total active power generation in the system matches the predicted power demand P_d plus the transmission losses:

$$P_d = \sum_{i=1}^{N_N} P_{Ni} + \sum_{i=1}^{N_T} P_{Ti} + \sum_{i=1}^{N_H} P_{Hi} - P_L \tag{4.2}$$

where P_{Hi} denotes the ith hydropower generation; the transmission losses P_L are expressed by the loss formula

$$P_L = \sum_{i=1}^{N_N} \sum_{j=1}^{N_N} P_{Ni} P_{Nj} b_{1ij} + 2 \sum_{i=1}^{N_N} \sum_{j=1}^{N_T} P_{Ni} P_{Tj} b_{2ij} + 2 \sum_{i=1}^{N_N} \sum_{j=1}^{N_H} P_{Ni} P_{Hj} b_{3ij} + \sum_{i=1}^{N_T} \sum_{j=1}^{N_T} P_{Ti} P_{Tj} b_{4ij} + 2 \sum_{i=1}^{N_T} \sum_{j=1}^{N_H} P_{Ti} P_{Hj} b_{5ij} + \sum_{i=1}^{N_H} \sum_{j=1}^{N_H} P_{Hi} P_{Hj} b_{6ij} \tag{4.3}$$

with the parameters b_{ij} having known values.

3. With the water transport delay between the $(i-1)$th and ith hydroplants assumed a given constant τ_{i-1} the reservoir's dynamics are described by

$$\mathring{S}_1(t) = i_1(t) - q_1(t) \tag{4.4a}$$

$$\mathring{S}_i(t) = i_i(t) - q_i(t) + q_{i-1}(t - \tau_{i-1}) \qquad i = 2, 3, \ldots, N_H \tag{4.4b}$$

where i_1 is the rate of natural water inflow to the ith reservoir. q_i is the ith water discharge rate.

4. The water volume at the ith reservoir is given by

$$S_i = d_{1i} + d_{2i}H_i + d_{3i}H_i^2 \tag{4.5}$$

This corresponds to trapezoidal reservoirs. H_i denotes the forebay elevation at the ith reservoir. The known parameters d_i are determined by the reservoir's dimensions (Ref. 4.10).

5. The ith hydroplant's power generation is given by the double quadratic formula

$$P_{Hi} = [\alpha_{1i} + \alpha_{2i}H_i + \alpha_{3i}H_i^2][b_{1i} + b_{2i}q_i + b_{3i}q_i^2] \tag{4.6}$$

This formula includes the effect of variable efficiency and variable head. The parameters α_i and b_i can be obrained by fitting equation (4.6) to the plant capability curves.

6. The volume of water discharge at any hydroplant is a specified constant C_i:

$$\int_0^T q_i(t)\, dt = C_i \tag{4.7}$$

7. Physical limitations impose the magnitude constraints

$$q_i^m \leq q_i \leq q_i^M \tag{4.8}$$

$$H_i^m \leq H_i \leq H_i^M \tag{4.9}$$

8. The transient changes of the average xenon 135 concentration in the ith core, which are induced by power level changes, are accounted for by the kinetic equations (Ref. 4.11)

$$\mathring{I}_i = -\lambda_I I_i + \frac{\gamma_I}{e_i G} P_{Ni} \tag{4.10}$$

$$\mathring{X}_i = -\lambda_X X_i + \lambda_I I_i - \frac{\Gamma_X}{e_i \sum_{f_i} G} X_i P_{Ni} + \frac{\gamma_X}{G e_i} P_{Ni} \tag{4.11}$$

where I_i and X_i denote the iodine 135 and xenon 135 average concentrations, respectively. λ_I and λ_X respresent the iodine and xenon radioactive decay constants. $\sum_{f_i}$ is the effective fission macroscopic cross section of the ith core. Γ_X denotes the xenon's thermal neutron microscopic absorption cross section. G is the energy released per fission. e_i is the plant's overall efficiency. γ_I and γ_X denote the iodine and xenon fission yield.

9. The effects of power level and xenon concentration on the reactivity ρ_i of the ith core are described by

$$\rho_i = \rho_{0i} - \rho_{Ci} - \alpha_{di}P_{Ni} - \alpha_{Xi}X_i \tag{4.12}$$

where ρ_{0i} is a constant reference reactivity. The known parameters α_{di} and α_{Xi} denote the power and xenon feedback coefficients (Ref. 4.5). ρ_{Ci} is the external reactivity provided by the control devices whose compensating action against variations of power level and xenon concentration maintains the reactor critical, that is $\rho_i = 0$.

10. Only limited xenon overrride capability is available in most reactors. This is accounted for by upper and lower magnitude constraints on the external control reactivity ρ_{Ci}. Without loss of geneality we assume

$$0 \leq \rho_{Ci} \leq \rho_{Ci}^M \tag{4.13}$$

11. Physical limitations impose the magnitude constraints

$$P_{Ni}^m \leq P_{Ni} \leq P_{Ni}^M \tag{4.14}$$

$$P_{Ti}^m \leq P_{Ti} \leq P_{Ti}^M \tag{4.15}$$

4.3. The Optimization Procedure

Here the optimization problem of Section 4.2 is reformulated as a minimum norm problem in a Hilbert space with quadratic norm. This formulation requires each of the system's variables to appear in the augmented cost functional raised to the second power (Ref. 4.12). With this in mind it is convenient (1) to eliminate from the system's equations the variables I_i, X_i, and S_i, which only appear linearly, (2) to introduce a change of variable in the case of ρ_{Ci}, which appears linearly and cannot be eliminated from the system's equations, and (3) to introduce pseudo-control-variables in order to eliminate those terms that are cubic or higher.

By differentiating equation (4.5) and substituting in (4.4) the variable S_i is eliminated. The resulting expression relates the forebay elevation and the water discharge as follows:

$$0 = q_i(t) - q_{i-1}(t - \tau_{i-1}) + d_{2i}\mathring{H}_i(t) + 2d_{3i}H_i(t)\mathring{H}_i(t) - i_i(t) \tag{4.16a}$$

$$0 = q_1(t) - i_1(t) + d_{21}\mathring{H}_1(t) + 2d_{31}\mathring{H}_1(t)H_1(t) \tag{4.16b}$$

From equations (4.10), (4.11), and (4.12) X_i and I_i can be eliminated. Making the change of variable

$$\xi_i^2 = \rho_{Ci} \tag{4.17}$$

and introducing the pseudo-control-variables ψ_i snd ϕ_i, where

$$\psi_i^2 = P_{Ni} \tag{4.18}$$

$$\phi_i = \psi_i \xi_i \tag{4.19}$$

the following expression relating the power generation and the external reactivity is obtained:

$$\begin{aligned} \mathring{P}_{Ni}(t) &= A_{1i}P_{Ni}(t) + A_{2i}P_{Ni}^2(t) + A_{3i}\phi_i^2(t) + g_i(t) \\ &\quad + A_{4i}\xi_i^2(t) + A_{5i}\xi_i\mathring{\xi}_i + A_{6i}\int_0^t e^{-\lambda_I(t-\tau)}P_{Ni}(\tau)\,d\tau \end{aligned} \tag{4.20}$$

where the coefficients A_{ji} and the functions $g_i(t)$ are known. The details of this derivation appear in Appendix 4.A.

With the help of the pseudo-control-variable y_i

$$y_i(t) = q_i(t)H_i(t) \tag{4.21}$$

the cubic terms in the hydrogeneration formula (4.6) are replaced with quadratic terms. After expanding (4.6) and substituting (4.21) the hydrogeneration formula is rewritten as follows:

$$\begin{aligned} P_{Hi} &= C_{1i} + C_{2i}H_i + C_{3i}H_i^2 + C_{4i}q_i + C_{5i}q_iH_i \\ &\quad + C_{6i}y_iH_i + C_{7i}q_i^2 + C_{8i}q_iy_i + C_{9i}y_i^2 \end{aligned} \tag{4.22}$$

where the coefficients C_{ji} are known.

The nonlinear equations (4.2), (4.16), (4.18), (4.19), (4.20), (4.21), and (4.22) and the magnitude constraints (4.8), (4.9), (4.13), (4.14), and (4.15) are included in the augmented cost functional by using unknown functions as follows:

a. The power balance equation (4.2) corresponds to the unknown function θ.
b. Equation (4.16) corresponds to the unknown functions M_i.
c. The pseudoconstraints (4.18) and (4.19) correspond to the functions Λ_i and Ω_i, respectively.
d. The nuclear generation equation (4.20) corresponds to R_i.
e. The pseudoconstraint (4.21) and the hydrogeneration formula (4.22) correspond to the unknown functions Z_i and T_i, respectively.

f. The right-hand sides of inequalities (4.8), (4.9), (4.13), (4.14), and (4.15) correspond to the Kuhn–Tucker multiplier functions η^u_{qi}, η^u_{Hi}, $\eta^u_{\rho i}$, η^u_{Ni}, and η^u_{Ti}, respectively.
g. The left-hand sides of inequalities (4.8), (4.9), (4.13), (4.14), and (4.15) correspond to the Kuhn–Tucker multiplier functions η^L_{qi}, η^L_{Hi}, $\eta^L_{\rho i}$, η^L_{Ni}, and η^L_{Ti}, respectively.

The augmented objective functional is now written in a compact quadratic form. This is done by first defining the control vector

$$U = \mathrm{col}[\boldsymbol{\phi}, \boldsymbol{\psi}, \boldsymbol{\xi}, P_{\mathbf{N}}, P_{\mathbf{T}}, P_{\mathbf{H}}, \mathbf{H}, \mathbf{y}, \mathbf{q}] \tag{4.23}$$

(where the notation $\mathbf{X}$ represents a column vector with elements x_i) and then eliminating all those terms that do not explicitly depend on U:

$$J(U) = \int_0^T [U^T F U + L^T U]\, dt \tag{4.24}$$

here F is a symmetric matrix, required by this formulation to be positive definite, and L is a column vector. The relations presented in Appendix 4.B are used through the derivation of (4.24).

The matrix F is partitioned in the form

$$F = \mathrm{diag}[F_1, F_2, F_3, F_4] \tag{4.25}$$

where F is a diagonal matrix of N_Nth order with elements

$$F_{1ii} = R_i A_{3i} \tag{4.26}$$

F_2 is a symmetric matrix of $2N_N$th order

$$F_2 = \begin{bmatrix} F_{21} & F_{22} \\ F_{22} & F_{23} \end{bmatrix} \tag{4.27}$$

where

$$F_{21} = \mathrm{diag}\{\Lambda_i\}, \qquad i = 1, 2, \ldots, N_N \tag{4.28}$$

$$F_{22} = \{-\Omega/2\} \tag{4.29}$$

and

$$F_{24} = \mathrm{diag}\{A_{4i} R_i - A_{5i} \mathring{R}_i\} \tag{4.30}$$

F_3 is a symmetric matrix of $(N_N + N_T + N_H)$th order partitioned as follows:

$$F_3 = \begin{bmatrix} F_{31} & F_{32} & F_{33} \\ F^T_{32} & F_{34} & F_{35} \\ F^T_{33} & F^T_{35} & F_{36} \end{bmatrix} \tag{4.31}$$

where F_{31} is a N_Nth-order symmetric matrix with elements

$$F_{31ij} = \begin{cases} R_iA_{2i} + \gamma_{Ni} - \theta b_{1ij}, & i = j \\ -\theta b_{1ij}, & i \neq j \end{cases} \tag{4.32}$$

F_{32} is a $N_N \times N_T$ rectangular matrix with elements

$$F_{32ij} = b_{2ij}\theta \tag{4.33}$$

F_{33} is a $N_N \times N_H$ rectangular matrix with elements

$$F_{33ij} = -b_{3ij}\theta \tag{4.34}$$

F_{34} is a symmetric matrix of order N_T, where

$$F_{34ij} = \begin{cases} \gamma_{Ti} - \theta b_{4ij} \\ -b_{4ij} \end{cases} \tag{4.35}$$

F_{35} is a $N_T \times N_H$ rectangular matrix with elements

$$F_{35ij} = -\theta b_{5ij} \tag{4.36}$$

F_{36} is a N_Hth-order symmetric matrix:

$$F_{36ij} = -\theta b_{6ij} \tag{4.37}$$

Finally, F_4 is a symmetric matrix of $3N_H$th order partitioned in the form

$$F_4 = \begin{bmatrix} F_{41} & F_{42} & F_{43} \\ F_{42} & F_{44} & F_{45} \\ F_{43} & F_{45} & F_{46} \end{bmatrix} \tag{4.38}$$

where every partition forms a N_Hth-order diagonal matrix:

$$F_{41} = \text{diag}\{T_iC_{3i} - d_{3i}\mathring{M}_i\} \tag{4.39}$$

$$F_{42} = \text{diag}\{\tfrac{1}{2}T_iC_{6i}\} \tag{4.40}$$

$$F_{43} = \text{diag}\{\tfrac{1}{2}C(T_iC_{5i} + Z_i)\} \tag{4.41}$$

$$F_{44} = \text{diag}\{T_iC_{9i}\} \tag{4.42}$$

$$F_{45} = \text{diag}\{\tfrac{1}{2}T_iC_{8i}\} \tag{4.43}$$

$$F_{46} = \text{diag}\{C_{7i}T_i\} \tag{4.44}$$

The column vector L is partitioned in the form

$$L = \text{col}[\mathbf{L}_1, \mathbf{L}_2, \ldots, \mathbf{L}_9] \tag{4.45}$$

where

$$L_{1i} = \Omega_i, \qquad i = 1, 2, \ldots, N_N \tag{4.46}$$

$$L_{2i} = 0, \qquad i = 1, 2, \ldots, N_N \tag{4.47}$$

$$L_{3i} = \eta^u_{\rho i} - \eta^L_{\rho i}, \qquad i = 1, 2, \ldots, N_N \tag{4.48}$$

$$L_{4i} = \beta_{Ni} + \mathring{R}_i + A_{1i}R_i + A_{6i}\int_t^T e^{\lambda_I(t-\tau)}R_i(\tau)\,d\tau$$

$$+ \theta + \eta^u_{Ni} - \eta^L_{Ni}, \qquad i = 1, 2, \ldots, N_N \tag{4.49}$$

$$L_{5i} = \beta_{Ti} + \theta + \eta^u_{Ti} - \eta^L_{Ti}, \qquad i = 1, 2, \ldots, N_T \tag{4.50}$$

$$L_{6i} = \theta - T_i, \qquad i = 1, 2, \ldots, N_H \tag{4.51}$$

$$L_{7i} = T_iC_{2i} - d_{2i}\mathring{M}_i + \eta^u_{Hi} - \eta^L_{Hi}, \qquad i = 1, 2, \ldots, N_H \tag{4.52}$$

$$L_{8i} = -Z_i, \qquad i = 1, 2, \ldots, N_H \tag{4.53}$$

$$L_{9i} = T_iC_{4i} + M_i - \tilde{\phi}(t, T, \tau_i)M_{i+1}(t + \tau_i)$$

$$+ \eta^u_{qi} - \eta^L_{qi}, \qquad i = 1, 2, \ldots, N_{H-1} \tag{4.54}$$

$$L_{9i} = T_iC_{4i} + M_i + \eta^u_{qi} - \eta^L_{qi}, \qquad i = N_H$$

here $\tilde{\phi}(t, T, \tau)$ denotes the function

$$\tilde{\phi}(t, T, \tau) = \begin{cases} 0, & t > T - \tau \\ 1, & t \leq T - \tau \end{cases} \tag{4.55}$$

The unknown functions R_i and M_i satisfy the homogeneous boundary conditions $R_i(T) = 0$ and $M_i(T) = 0$.

The problem of Section 4.2 is reformulated now as minimizing the objective functional (4.24) subject to the linear constraints (4.7). For convenience equation (4.7) is expressed in terms of U:

$$\mathbf{C} = \int_0^T EU\,dt \tag{4.56}$$

where E is a $N_H \times (4N_N + 4N_H + N_T)$ rectangular matrix partitioned as follows:

$$E = [0|I] \tag{4.57}$$

with I being the N_Hth-order identity matrix.

4.4. The Optimal Solution

It has been shown (Ref. 4.8) that the problem of minimizing an objective functional in the form of (4.24) subject to a linear constraint can be reduced to the following minimum norm problem in the framework of functional analysis (Ref. 4.9).

Let B and D be Banach spaces and T a bounded linear transformation defined on B with values in D. For each α in the range of T and for $\hat{\mu} \in B$, the unique element $\mu_\alpha \in B$ that satisfies

$$\alpha = T\mu \tag{4.58}$$

while minimizing the norm

$$J(\mu) = \|\mu - \hat{\mu}\| \tag{4.59}$$

is given by

$$\mu_\alpha = T^+[\alpha - T\hat{\mu}] + \hat{\mu} \tag{4.60}$$

where T^+ denotes the pseudoinverse of T.

Here the vector U defined in (4.23) is associated with the element μ of the minimum norm problem, the objective functional (4.24) corresponds to the norm (4.59). The linear constraint (4.56) corresponds to (4.58) and the vector $-\frac{1}{2}F^{-1}L$ is associated with $\hat{\mu}$.

The solution to our problem is therefore given by (4.60) in the form

$$U_{\text{opt}} = F^{-1}E^T\left[\int_0^T EF^{-1}E^T d_t\right]\left[\mathbf{C} + \tfrac{1}{2}\int_0^T EF^{-1}L\,dt\right] - \tfrac{1}{2}F^{-1}L \tag{4.61}$$

Substituting F, E, and L in (4.61) the conditions for optimality are obtained:

a. $\Omega_i = -2R_i\phi_i A_{3i}$ (4.62)

b. $\Omega_i\xi_i - 2\Lambda_i\psi_i = 0$ (4.63)

c. $\Omega_i\psi_i - 2A_{4i}R_i\xi_i + 2A_{5i}\xi_i\mathring{R}_i - \eta^u_{\rho i} + \eta^L_{\rho i} = 0$ (4.64)

d. $$\mathring{R}_i = -(A_{1i} + 2A_{2i}P_{Ni})R_i - A_{6i}\int_t^T e^{\lambda_I(t-\tau)} \times R_i(\tau)d\tau - 2\gamma_{Ni}P_{Ni} + 2\theta\sum_{j=1}^{N_N} b_{1ij}P_{Nj} - \theta - \eta^u_{Ni} + \eta^L_{Ni} + 2\theta\sum_{j=1}^{N_T} b_{2ij}P_{Tj} + 2\theta\sum_{j=1}^{N_H} b_{3ij}P_{Hj} \tag{4.65}$$

with $R_i(T) = 0$

e. $$2\theta\left[\sum_{j=1}^{N_N} b_{2ji}P_{Nj} + \sum_{j=1}^{N_T} b_{4ij}P_{Tj} + \sum_{j=1}^{N_H} b_{5ij}P_{Hj}\right] + \eta^u_{Ti} - \eta^L_{Ti} - \partial\gamma_{Ti}P_{Ti} - \beta_{Ti} - \theta = 0 \tag{4.66}$$

f. $$T_i = \left[1 - 2\left(\sum_{j=1}^{N_N} b_{3ji}P_{Nj} + \sum_{j=1}^{N_T} b_{5ji}P_{Tj} + \sum_{j=1}^{N_H} b_{6ij}P_{Hj}\right)\right]\theta \tag{4.67}$$

g. $$Z_i = T_iC_{6i}H_i + 2T_i(q_iy_i) + T_iC_{8i}q_i \tag{4.68}$$

h. $$\mathring{M}_i(2H_id_{3i} + d_{2i}) = T_i(C_{2i} + 2C_{3i}H_i + C_{6i}y_i + C_{5i}q_i) + \eta^u_H - \eta^L_{Hi} + Z_iq_i \tag{4.69}$$

with $M_i(T) = 0$

i. $$q_i = G_{6i}\left[\int_0^T G_{6i}\,dt\right]^{-1}\left[C_i + \tfrac{1}{2}\int_0^T (G_{3i}L_{7i} + G_{5i}L_{8i} + G_{6i}L_{qi})\right] - \tfrac{1}{2}[G_{3i}L_{7i} + G_{5i}L_{8i} + G_{6i}L_{qi}] \tag{4.70}$$

where

$$G_{3i} = \frac{T_i^2(\frac{1}{4}C_{6i}C_{8i} - C_{5i}C_{9i}\frac{1}{2}) - \frac{1}{2}T_iZ_iC_{9i}}{\Delta_i} \tag{4.71}$$

$$\begin{aligned}\Delta_i = {} & T_i^3(C_{3i}C_{9i}C_{7i} + \tfrac{1}{4}C_{6i}C_{5i}C_{8i} \\ & - \tfrac{1}{4}C_{5i}^2C_{9i} - \tfrac{1}{4}C_{8i}^2C_{3i} - \tfrac{1}{4}C_{6i}^2C_{7i}) \\ & + T_i^2\mathring{M}_i(\tfrac{1}{4}C_{8i}^2d_{3i} - C_{9i}C_{7i}d_{3i}) \\ & + T_i^2Z_i(\tfrac{1}{4}C_{6i}C_{8i} - \tfrac{1}{2}C_{5i}C_{9i}) \\ & - Z_i^2T_i(\tfrac{1}{4}C_{9i})\end{aligned} \tag{4.72}$$

$$G_{5i} = [T_i^2(\tfrac{1}{4}C_{6i}C_{5i} - \tfrac{1}{2}T_i^2C_{8i}C_{3i}) + \tfrac{1}{4}T_iZ_iC_{6i} + \tfrac{1}{2}d_{3i}C_{8i}\mathring{M}_iT_i]/\Delta_i \tag{4.73}$$

$$G_{6i} = [T_i^2(C_{3i}C_{9i} - \tfrac{1}{4}C_{6i}^2) - T_iC_{9i}d_{3i}\mathring{M}_i]/\Delta_i \tag{4.74}$$

The Kuhn–Tucker multiplier functions η_i^u and η_i^L are obtained such that the exclusion equations associated with the inequality constraints are satisfied.

The conditions for optimality can be further simplified by solving for Ω_i, Λ_i, ψ_i, and ϕ_i from equations (4.18), (4.19), (4.62), and (4.63) and then substituting in (4.64) in order to solve for ξ_i. Doing this the following expression is obtained:

$$\xi_i = \frac{(\eta^u_\rho - \eta^L_{\rho i})}{2[A_{5i}\mathring{R}_i - (A_{3i}P_{Ni} + A_{4i})R_i]} \tag{4.75}$$

Using (4.75) in conjunction with the exclusion equations

$$\eta^u_{\rho i}[\xi_i - (\rho_i^M)^{1/2}] = 0, \qquad \eta^u_{\rho i} \geq 0 \tag{4.76}$$

and

$$\eta^L_{\rho i}[-\xi_i] = 0, \qquad \eta^L_{\rho i} \geq 0 \tag{4.77}$$

η_{ρ}^{L} and $\eta_{\rho i}^{u}$ are eliminated yielding the final result:

$$\xi_i = \begin{cases} (\rho_i^M)^{1/2} & \text{if } A_{5i}\mathring{R}_i - (A_{3i}P_{Ni} - A_{4i})R_i > 0 \\ 0 & \text{if } A_{5i}\mathring{R}_i - (A_{3i}P_{Ni} - A_{4i})R_i < 0 \end{cases} \tag{4.78}$$

4.5. A Feasible Multilevel Approach

A two-level iterative approach is proposed for implementing the optimal solution (Ref. 4.13). We ensure that at each iteration step between levels the system constraints are satisfied. Therefore every iteration constitutes a feasible suboptimal solution.

The iterative scheme is composed of two levels with each level divided into three sublevels as shown in Fig. 4.1. The variables attached to arriving arrows are treated at each level as known functions and only those variables appearing inside the blocks are treated as unknown. Here an iteration cycle is described.

Given an initial guess $U_0(t)$, θ is computed at sublevel (2b) from the Kth equation (4.66):

$$\theta(t) = \frac{(2\gamma_{TK}P_{TK} + \beta_{TK} + \eta_{TK}^{L} - \eta_{TK}^{u})}{2\left(\sum_{j=1}^{N_N} b_{2jk}P_{nj} + \sum_{j=1}^{N_T} b_{4kj}P_{Tj} + \sum_{j=1}^{N_H} b_{5kj}P_{Hj} - \frac{1}{2}\right)} \tag{4.79}$$

This result is then fed into sublevels (2a) and (2c). At sublevel (2a) $R_i(t)$ is computed by solving equation (4.65) backward in time.

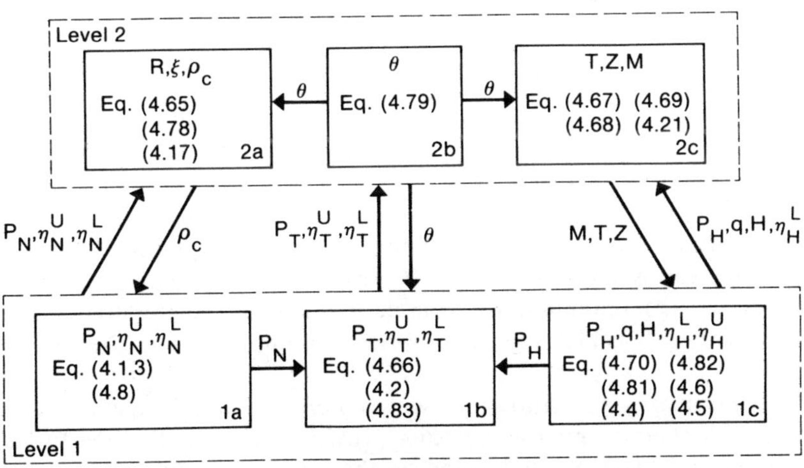

Figure 4.1. A two-level iteration scheme.

ξ_i and ρ_{Ci} are obtained from equations (4.78) and (4.17), respectively. At sublevel (2c) the variables T_i, Z_i, and M_i are obtained from equations (4.67), (4.68), (4.69), and (4.21). M_i is computed by solving (4.69) backward in time.

The information obtained at level 2 is then fed back into level 1 as shown in the figure. At sublevel (1a) the nuclear power generations are obtained from equation (4A.3) in Appendix 4.A and the exclusion equations

$$\begin{aligned} \eta^u_{Ni}[P_{Ni} - P^M_{Ni}] &= 0, \qquad \eta^u_{Ni} \geq 0 \\ \eta^L_{Ni}[P^m_{Ni} - P_{Ni}] &= 0, \qquad \eta^L_{Ni} \geq 0 \end{aligned} \tag{4.80}$$

At sublevel (1c) the variables q_i are obtained from equation (4.70) and the exclusion equations

$$\begin{aligned} \eta^u_{qi}[q_i - q^M_i] &= 0, \qquad \eta^u_{qi} \geq 0 \\ \eta^L_{qi}[q^m_i - q_i] &= 0, \qquad \eta^L_{qi} \geq 0 \end{aligned} \tag{4.81}$$

The forebay elevations H_i are obtained by solving equations (4.4) and (4.5) and the exclusion equations

$$\begin{aligned} \eta^u_{Hi}[H_i - H^M_i] &= 0, \qquad \eta^u_{Hi} \geq 0 \\ \eta^L_{Hi}[H^m_i - H_i] &= 0, \qquad \eta^L_{Hi} \geq 0 \end{aligned} \tag{4.82}$$

The hydropower generations P_{Hi} are given by equation (4.6).

Finally the fossil plant generations are computed at level (1b) by solving simultaneously the remaining $N_T - 1$ equations (4.66), the power balance equation (4.2), and the exclusion equations

$$\begin{aligned} \eta^u_{Ti}[P_{Ti} - P^M_{Ti}] &= 0, \qquad \eta^u_{T_i} \geq 0 \\ \eta^L_{Ti}[P^m_{Ti} - P_{Ti}] &= 0, \qquad \eta^L_{Ti} \geq 0 \end{aligned} \tag{4.83}$$

4.6. Conclusions and Comments

The optimal solution to the problem of short-term operation of hydro-thermal–nuclear systems is obtained by use of an optimization technique that employs the minimum norm formulation. A two-level iterative scheme of the so-called feasible type is proposed for implementing the optimal solution.

The optimality conditions presented here are applicable to power systems with thermal nuclear reactors provided with continuous fueling capabilities and hydroplants located on the same stream with trapezoidal reservoirs.

The nuclear model considered is only a crude approximation to the very complex processes that take place in a nuclear reactor core. It would be impractical, if not impossible, to consider in an application of this nature a very detailed distributed model describing the reactor core processes. It is thought that the present methodology can be applied to more sophisticated (but still relatively simple) reactor models where the spatial effects in the core are considered.

Appendix 4.A

For a critical reactor the reactivity ρ_i is zero. Solving for x_i from (4.12) we obtain

$$x_i = \frac{\rho_{0i}}{\alpha_{xi}} - \frac{\rho_{ci}}{\alpha_{xi}} - \frac{\alpha_{di}}{\alpha_{xi}} P_{Ni} \tag{4A.1}$$

From (4.10), assuming an initial condition $I_i(t_0) = I_{0i}$,

$$I_i(t) = I_{0i} e^{-\lambda_I(t-t_0)} + \frac{\gamma_I}{e_i G} \int_0^t e^{-\lambda_I(t-\tau)} P_{Ni}(\tau)\, d\tau \tag{4A.2}$$

Differentiating (4A.1) and substituting both (4A.1) and (4A.2) in equation (4.11) gives

$$\begin{aligned} \mathring{P}_{Ni} = {} & \left(-\lambda_X + \frac{\Gamma_x \rho_{0i}}{e_i \sum_{fi} G\alpha_{di}} - \frac{\gamma_X \alpha_{xi}}{e_i G \alpha_{di}}\right) P_{Ni} \\ & - \frac{\Gamma_x}{e_i \sum_{fi} G} P_{Ni}^2 - \frac{\Gamma_x}{e_i \sum_{fi} G\alpha_{di}} P_{Ci} P_{Ni} \\ & + \left(\frac{\lambda_x \rho_{0i}}{\alpha_{di}} - \frac{\lambda_I \alpha_{xi}}{\alpha_{di}} I_0\, e^{-\lambda_I(t-t_0)}\right) - \frac{\lambda_X}{\alpha_{di}} \rho_i \\ & - \frac{\mathring{\rho}_{Ci}}{\alpha_{di}} - \frac{\lambda_I \alpha_{xi} \gamma_I}{\alpha_{di} e_i G} \int_0^t e^{-\lambda_I(t-\tau)} P_{Ni}(\tau)\, d\tau \end{aligned} \tag{4A.3}$$

Appendix 4.B

The following relations are used in the derivation of the objective functional (4.24):

a. $$2\int_0^T \mu(t) v(t) \mathring{v}(t)\, dt = \mu(T) v^2(T) - \mu_0 v_0^2 - \int_0^T v^2(t) \mathring{\mu}(t)\, dt$$

b. $$\int_0^T \mu(t)V(t-\tau)\,dt$$

$$= \int_{-\tau}^{0} \mu(t-\tau)V(t)\,dt + \int_0^T \tilde{\phi}(t, T, \tau)\mu(t+\tau)V(t)\,dt$$

where

$$\tilde{\phi}(t, T, \tau) = \begin{cases} 1, & t \leq T - \tau \\ 0, & t > T - \tau \end{cases}$$

c. $$\int_0^T \int_0^t \mu(t)\, e^{-\lambda_l(t-\lambda)} V(\tau)\, d\tau\, dt = \int_0^T \int_t^T V(t)\, e^{\lambda_l(t-\tau)} \mu(\tau)\, d\tau\, dt$$

References

4.1. KIEFER, W. M., and KONCEL, E. F., "Scheduling Generations on Systems with Fossil and Nuclear Units," *Trans. Am. Nucl. Soc.* **13**, 768 (1970).

4.2. HOSKINS, R. E., and REES, F. J., "Power Systems Optimization Approach to Nuclear Fuel Management," *Trans. Am. Nucl. Soc.* **13**, 768 (1970).

4.3. GROSSMAN, L. M., and REINKING, A. G., "Fuel Management and Load Optimization of Nuclear Units in Electric Systems," *Trans. Am. Nucl. Soc.* **20**, 391 (1975).

4.4. CHOU, Q. B., "Characteristics and Maneuverability of Candu Nuclear Power Stations Operated for Base-Load and Load Following Generation," *IEEE Trans. Power Appar. Syst.* **PAS-94**(3), May/June, 792–801 (1975).

4.5. EL-WAKIL, M. M., *Nuclear Power Engineering*, McGraw-Hill, New York, 1962.

4.6. YASUKAWA, S., "An Analysis of Continuous Reactor Refueling," *Nucl. Sci. Eng.* **24**, 253–260 (1966).

4.7. MILLAR, C. H., "Fuel Management in Candu Reactors," *Trans. Am. Nucl. Soc.* **20**, 350 (1975).

4.8. EL-HAWARY, M. E., and CHRISTENSEN, G. S., *Optimal Economic Operation of Electric Power Systems*, Academic, New York, 1979.

4.9. PORTER, W. A., *Modern Foundations of Systems Engineering*, Macmillan, New York, 1966.

4.10. HAMILTON, E. P., and LAMONT, I. W., "An Improved Short Term Hydro-Thermal Coordination Model," Paper No. A77 518-4, Institute of Electrical and Electronics Engineers Summer Power Meeting, Mexico City, 1977.

4.11. ISBIN, H. S., *Introductory Nuclear Reactor Theory*, Reinhold, New York, 1963.

4.12. SHAMALY, A., *et al.*, "A Transformation for Necessary Optimality Conditions for Systems with Polynomial Nonlinearities," *IEEE Trans. Autom. Control* **AC-24**, 983–985 (1979).

4.13. MAHMOUD, M. S., "Multilevel Systems Control and Applications: A Survey," *IEEE Trans. Syst. Man Cybern.* **7**(3), 125–143 (1977).

4.14. NIEVA, R., CHRISTENSEN, G. S., and EL-HAWARY, M. E., "Functional Optimization of Nuclear-Hydro-Thermal Systems," Proceedings, CEC, Toronto, 1978.

4.15. NIEVA, R., CHRISTENSEN, G. S., and EL-HAWARY, M. E., "Optimum Load Scheduling of Nuclear-Hydro-Thermal Power Systems," *Optimization Theory Appl.* **35**(2), 261–275 (1981).

5

Optimal Tie-Line Control

5.1. Introduction

For more than two decades, the problem of load frequency control (LFC) using conventional or advanced-control theory has been the subject of numerous studies. The conventional LFC approach often employs what is called the tie-line bias concept to design a system controller that has a proportional-plus-integral (PI) action. This type of control is used extensively in practice in preference to all the techniques that have been proposed recently, employing modern control theory. The reason for that is that most recent techniques have developed linear feedback controls that are functions of all the system state variables as well as the system disturbances (Refs. 5.1, 5.2). Therefore, it was necessary to design an observer to realize these kind of controls (Ref. 5.3). Once an observer is introduced into the system, the cost is increased, and the control is no longer optimal (Ref. 5.4). Another important reason is that a control that depends upon all the system states needs some of these state variables to be telemetered, since the areas of interconnected power systems (IPS) are spread over large geographical territories. This is why, in practice, control engineers prefer to use the conventional control to the advanced one, in spite of the contention that the latter improves the system transient performance.

Calovic (Refs. 5.5, 5.6) was one of the first to narrow the gap between the conventional LFC and the advanced LFC by adopting Porter's idea (Ref. 5.7) of introducing the integral of the output states as a part of the optimum control design. However, the proportional part of the control is a function of all the state variables, and an observer is necessary to realize the control law. Calovic (Ref. 5.8) has developed a PI control law in which the proportional part as well as the integral part is only a function of the output state variables. The necessary condition to validate this control depends upon the choice of a certain weighting matrix.

There is, however, no systematic way of finding this weighting matrix. Work reported in this chapter implements modern control theory to find the optimal parameters of the proportional and the integral parts of the control law in a systematic way so as to meet the system transient and steady-state specifications, as well as to meet the kind of control that engineers exercise in the real world when controlling the generation in power systems.

In this chapter the solution of the optimal tie-line control problem is discussed. The first section is concerned with the design of a linear regulator for the problem of LFC of IPS. The control is specified to be of PI form and is only a function of the measurable states. The LFC problem is formulated as a parameter optimization problem by employing the Pontryagin minimum principle. The second section of this chapter is concerned with the optimum LFC of multiarea interconnected power systems. The control law in that section is assumed to have a proportional and/or integral form and is a function only of the measurable states (present-day practice). The third section is concerned with the design of an optimal controller for a system whose dynamical equations contain a backlash element; this design is applied to the problem of LFC of a single area steam power system.

5.2. Load Frequency Control of Interconnected Power Systems

The case considered in this section is that of a two mixed area interconnected power system (TMAIPS), which may be regarded as a representation of a particular area, which undergoes a disturbance, connected, via a tie line, to the rest of the system. The disturbed area is assumed to be subjected to a step load change ΔL. The justification for assuming that IPS are subjected to deterministic load disturbances, in spite of the fact that they are normally subject to random disturbances, is that modern LFC systems are designed with filters (Ref. 5.9) in order to remove the purely random portion of the regulating responsibility, which generation cannot follow, leaving the deterministic components for the system units to follow.

Once an area has experienced a load change (e.g., load increase), a chain of events ensues as follows: (1) the disturbed area speed begins to decrease as the increased load demand is supplied from the stored kinetic energy; (2) the tie-line phase angle increases, and unscheduled tie-line power flows to the disturbed area; (3) the remote area speed begins to decrease, as the disturbed area unscheduled tie-line power represents a remote area load increase; (4) the governors on both areas sense the change in the speed, and each area responds in proportion to its natural governing characteristic; and (5) the supplementary regulators come into action either after the governors have ceased operating or during their operation.

The approach that will be adopted in designing the supplementary regulators depends on whether they come into action before or after the governors have stopped operating. LFC operation in which the governors cease operating before the action of the supplementary regulators will be referred to as noninteractive governor–supplementary regulator LFC. The case in which the supplementary regulators come into action during the operation of the governors will be referred to be as interactive governor–supplementary regulator LFC. The case considered in this book is the case of noninteractive governor–supplementary regulator action.

5.2.1. Mathematical Model

A linearized model of a TMAIPS can be written as (Ref. 5.3)

$$\dot{X}_1 = A_1 X^1 + B_1 U + \Gamma_1 \tag{5.1}$$

$$Y = C_1 X^1 \tag{5.2}$$

Here, the matrices A_1, B_1, Γ_1, C_1 are given by

$$A_1 = \begin{bmatrix} -G_1/M_1 & 0 & -1/M_1 & 0 & 0 & 1/M_1 & 0 & 0 \\ 0 & -G_2/M_2 & 1/M_1 & 0 & 0 & 0 & 0 & 1/M_2 \\ T_{12} & -T_{12} & 0 & 0 & 0 & 0 & 0 & 0 \\ -E_1/T_{g1} & 0 & 0 & -1/T_{g1} & 0 & 0 & 0 & 0 \\ 0 & -E_2/T_{g2} & 0 & 0 & -1/T_{g2} & 0 & 0 & 0 \\ 0 & 0 & 0 & 1/T_{t1} & 0 & -1/T_{t1} & 0 & 0 \\ 0 & 0 & 0 & 0 & 1/T_{t2} & 0 & -1/T_{t2} & 0 \\ 0 & 0 & 0 & 0 & -2/D_2 & 0 & 2/T_{t2}+2/D_2 & -2/D_2 \end{bmatrix} \tag{5.3}$$

$$B_1^T = \begin{bmatrix} 0 & 0 & 0 & -1/T_{g1} & 0 & 0 & 0 & 0 \\ 0 & 0 & 0 & 0 & -1/T_{g2} & 0 & 0 & 0 \end{bmatrix} \tag{5.4}$$

$$\Gamma_1^T = \begin{bmatrix} -\Delta_1/M_1 & 0 & 0 & 0 & 0 & 0 & 0 & 0 \\ 0 & -\Delta_2/M_2 & 0 & 0 & 0 & 0 & 0 & 0 \end{bmatrix} \tag{5.5}$$

$$C_1 = \begin{bmatrix} 1 & 0 & 0 & 0 & 0 & 0 & 0 & 0 \\ 0 & 1 & 0 & 0 & 0 & 0 & 0 & 0 \\ 0 & 0 & 1 & 0 & 0 & 0 & 0 & 0 \end{bmatrix} \tag{5.6}$$

The symbols are as follows: x_1^1, frequency deviation of area 1; x_2^1, frequency deviation of area 2; x_3^1, power angle deviation; x_4^1, deviation in the governor position of area 1; x_5^1, deviation in the governor position of area 2; x_6^1, deviation in mechanical output of area 1; x_7^1, deviation in the gate position of area 2; x_8^1, deviation in mechanical output of area 2; also, u_1, area 1 speed changer control command; u_2, area 2 speed changer control command.

5.2.2. Problem Formulation

Considering a noninteractive supplementary regulator action, the frequency of the interconnected system settles down to a new frequency common to all the areas of the system and to a new tie-line flow before the area's supplementary regulators come into action. For example, if area 1 is subjected to a load increase ΔL_1, the new system frequency is given by (Ref. 5.10)

$$f = f^s + -\Delta L_1/2\pi[(G_1 + E_1) + (G_2 + E_2)] \tag{5.7}$$

and the new tie-line power flow is

$$P_{t_{12}} = P^s_{t_{12}} + -\Delta L_1(G_2 + E_2)/(G_1 + E_1) + (G_2 + E_2) \tag{5.8}$$

Here, f^s is the scheduled frequency and $P^s_{t_{12}}$ is the scheduled tie-line power.

It is desired to design an optimal supplementary regulator with PI control of the area control error (ace), such that the system frequency and the tie-line power are maintained at their scheduled values. The ace is a measure of the prevailing load-generation mismatching. In terms of the system state variables, the control law can be written as

$$U(t) = -K_P \operatorname{ace}(t) - K_I \int_{t_0}^{t_f} \operatorname{ace}(t)\, dt \tag{5.9}$$

in which

$$\operatorname{ace}(t) = D_1 X^1(t) \tag{5.10}$$

$$K_P = \operatorname{diag}(k_{P1}, k_{P2}), \qquad K_I = \operatorname{diag}(k_{I1}, k_{I2}) \tag{5.11}$$

and D_1 denotes the matrix

$$D_1 = \begin{bmatrix} \beta_1 & 0 & 1 & 0 & 0 & 0 & 0 & 0 \\ 0 & \beta_2 & -1 & 0 & 0 & 0 & 0 & 0 \end{bmatrix} \tag{5.12}$$

The control parameters K_P and K_I can be expressed mathematically as follows:

$$\dot{k}_{Pi} = 0, \qquad i = 1, 2 \tag{5.13}$$

$$\dot{k}_{Ii} = 0, \qquad i = 1, 2 \tag{5.14}$$

The problem posed is to find the optimal control parameters k_{Pi} and k_{Ii}, $i = 1, 2$, which minimize the system transient and the control action such that the system steady state, dynamic limit, and area decentralization are met.

Minimization of the system transient and control effort can be accomplished by minimizing the cost functional

$$J_1 = (1/2) \int_{t_0}^{t_f} (X^{1T} Q_1 X^1 + U^T R U)\, dt \tag{5.15}$$

with respect to the control parameters K_P and K_I, subject to the dynamical constraints given by equations (5.1). Here, Q_1 and R are diagonal, positive-definite matrices of dimension 8 and 2, respectively.

The dynamic requirement is met when the steam area generation limit is kept within ψ p.u. MW/s; the generation rate limit is not fixed and varies from unit to unit ($60\psi = 2$–5). It can be expressed mathematically in terms of the system state variables as follows:

$$F_1^T X^1 \leq \psi \tag{5.16}$$

in which

$$F_1^T = [0, 0, 0, 1/T_{t1}, 0, -1/T_{t1}, 0, 0] \tag{5.17}$$

Ideally, to ensure the nonintervention principle of the interconnected load frequency control operation (that is, power generation must be altered only in the disturbed area), area frequency bias setting equal to the area natural governing characteristic must be employed in the ace equation (Ref. 5.13). When this condition is fulfilled, in the case of employing noninteractive governor–supplementary regulator, the control error of the remote area is equal to zero and its supplementary regulator remains inoperative.

To find the optimal control parameters K_I and K_P, the system dynamics given by equation (5.1) can be written as

$$\dot{X}^1 = A_1 X^1 - B_1 K_P D X^1 - B_1 K_I D \int_{t_0}^{t_f} X^1\, dt + \Gamma_1 \tag{5.18}$$

Equation (5.18) is the direct substitution of equation (5.4) into equation (5.1). Now, define a new variable Z by

$$Z = D_1 \int_{t_0}^{t_f} X^1\, dt \tag{5.19}$$

Then, define an augmented vector X such that

$$x_i = x_i^1, \qquad i = 1, \ldots, 8 \tag{5.20}$$

$$x_{8+j} = z_j, \qquad j = 1, 2 \tag{5.21}$$

Hence, the augmented dynamic system and the cost functional can be written as

$$\dot{X} = AX + \Gamma \tag{5.22}$$

$$J = (1/2) \int_{t_0}^{t_f} X^T Q X \, dt \tag{5.23}$$

Here,

$$A = \begin{bmatrix} A_1 - B_1 K_P D_1 & -B_1 K_I \\ D_1 & 0 \end{bmatrix} \tag{5.24}$$

0 being the 2×2 matrix whose elements are all zeros,

$$\Gamma^T = [\Gamma_1^T, \quad 0, \quad 0] \tag{5.25}$$

$$Q = \begin{bmatrix} Q_1 + D^T K_P^T R K_P D_1 & D_1^T K_P^T R K_I \\ K_I^T R K_P D & Q_2 + K_I^T R K_I \end{bmatrix} \tag{5.26}$$

Q_2 being a diagonal, positive-definite matrix with dimension 2.

The inequality constraints given in equation (5.16) can be written as

$$F^T X \leq \psi \tag{5.27}$$

in which

$$F^T = [F_1^T, \quad 0, \quad 0] \tag{5.28}$$

5.2.3. Optimal Solution

The inequality constraints given by equation (5.27) can be converted into equality constraints (Ref. 5.11) by defining a new variable x_{11} such that

$$\dot{x}_{11} = (FX)^2 H_1(FX) \tag{5.29}$$

$$H_1(FX) = \begin{cases} 0, & FX \leq 0 \\ 1, & FX > 0 \end{cases} \tag{5.30}$$

To handle a criterion of the form (5.23), the system differential equations and the constraint are augmented by an additional equation as follows:

$$\dot{x}_{12} = (1/2) X^T Q X \tag{5.31}$$

The Hamiltonian is defined as follows:

$$H = \sum_{i=1}^{12} \lambda_i \dot{x}_i \tag{5.32}$$

Then, by applying Pontryagin's minimum principle, one can get the system canonical equations. In the case of employing PI control action, the system dynamics can be written as follows:

$$\dot{x}_1 = -(G_1/M_1)x_1 - (1/M_1)x_3 + (1/M_1)x_6 - \Delta L_1/M_1 \tag{5.33}$$

$$\dot{x}_2 = -(G_2/M_2)x_2 + (1/M_2)x_3 + (1/M_2)x_8 - \Delta L_2/M_2 \tag{5.34}$$

$$\dot{x}_3 = T_{12}x_1 - T_{12}x_2 \tag{5.35}$$

$$\begin{aligned}\dot{x}_4 = &-(E_1/T_{g1})x_1 - (1/T_{g1})x_4\\ &-(\beta_1 k_{P1}/T_{g1})x_1 - (k_{P1}/T_{g1})x_3 - (k_{I1}/T_{g1})x_9\end{aligned} \tag{5.36}$$

$$\begin{aligned}\dot{x}_5 = &-(E_2/T_{g2})x_2 - (1/T_{g2})x_5\\ &-(\beta_2 k_{P2}/T_{g2})x_2 + (k_{P2}/T_{g2})x_3 - (k_{I2}T_{g2})x_{10}\end{aligned} \tag{5.37}$$

$$\dot{x}_6 = (1/T_{t1})x_4 - (1/T_{t2})x_6 \tag{5.38}$$

$$\dot{x}_7 = (1/T_{t2})x_5 - (1/T_{t2})x_7 \tag{5.39}$$

$$\dot{x}_8 = -(2/T_{t2})x_5 + [2(T_{t2} + D_2)]x_7 - (1/D_2)x_8 \tag{5.40}$$

$$\dot{x}_9 = \beta_1 x_1 + x_3 \tag{5.41}$$

$$\dot{x}_{10} = \beta_2 x_2 - x_3 \tag{5.42}$$

$$\dot{x}_{11} = [(1/T_{t1})x_4 - (1/T_{t1})x_6 - \psi]^2 H_1[(1/T_{t1})x_4 - (1/T_{t1})x_6 - \psi] \tag{5.43}$$

$$\begin{aligned}\dot{x}_{12} = &(1/2)\sum_{i=1}^{10} q_i x_i^2 + (1/2)r_1[k_{P1}(\beta_1 x_1 + x_3) + k_{I1}x_9]^2\\ &+ (1/2)r_2[k_{P2}(\beta_2 x_2 - x_3) + k_{I2}x_{10}]^2\end{aligned} \tag{5.44}$$

For example, in the case where area 1 is being subjected to a load change ΔL_1, the state initial conditions will be given as follows:

$$x_1(t_0) = x_2(t_0) = -\Delta L_1/(E_1 + G_1 + E_2 + G_2) \tag{5.45}$$

$$x_3(t_0) = -\Delta L_1(E_2 + G_2)/(E_1 + G_1 + E_2 + G_2) \tag{5.46}$$

$$x_4(t_0) = x_6(t_0) = -E_1 x_1(t_0) \tag{5.47}$$

$$x_5(t_0) = x_7(t_0) = x_8(t_0) = -E_2 x_2(t_0) \tag{5.48}$$

$$x_i(t_0) = 0, \qquad i = 9, 10, \ldots, 12 \tag{5.49}$$

The system costate equations can be written as

$$\dot{\lambda}_1 = -q_1x_1 - r_1\beta_1k_{P1}(\beta_1k_{P1}x_1 + k_{P1}x_3 + k_{I1}x_9) + (G_1/M_1)\lambda_1 - T_{12}\lambda_3 + [(E_1 + \beta_1k_{P1})/T_{g1}]\lambda_4 - \beta_1\lambda_9 \tag{5.50}$$

$$\dot{\lambda}_2 = -q_2x_2 - r_2\beta_2k_{P2}(\beta_2k_{P2}x_2 - k_{P2}x_3 + k_{12}x_{10}) + (G_2/M_2)\lambda_2 + T_{12}\lambda_3 + [(E_2 + \beta_2k_{P2})/T_{g2}]\lambda_5 - \beta_2\lambda_{10} \tag{5.51}$$

$$\dot{\lambda}_3 = -q_3x_3 - r_1k_{P1}(\beta_1k_{P1}x_1 + k_{P1}x_3 + k_{I1}x_9) + r_2k_{P2}(\beta_2k_{P2}x_2 - k_{P2}x_3 + k_{I12}x_{10}) + (1/M_1)\lambda_1 - (1/M_2)\lambda_2 + (k_{P1}/T_{g1})\lambda_4 - (k_{P2}/T_{g2})\lambda_5 - \lambda_9 + \lambda_{10} \tag{5.52}$$

$$\dot{\lambda}_4 = -q_4x_4 + (1/T_{g1})\lambda_4 - (1/T_{t1})\lambda_6 - (2/T_{t1}) \times [(1/T_{t1})x_4 - (1/T_{t1})x_6 - \psi]H_1\lambda_{11} \tag{5.53}$$

$$\dot{\lambda}_5 = -q_5x_5 + (1/T_{g2})\lambda_5 - (1/T_{t2})\lambda_7 + (2/T_{t2})\lambda_8 \tag{5.54}$$

$$\dot{\lambda}_6 = -q_6x_6 - (1/M_1)\lambda_1 + (1/T_{t1})\lambda_6 + (2/T_{t1}) \times [(1/T_1)x_4 - (1/T_{t1})x_6 - \psi]H_1\lambda_{11} \tag{5.55}$$

$$\dot{\lambda}_7 = -q_7x_7 + (1/T_{t2})\lambda_7 - (2/T_{t2} + 2/D_2)\lambda_8 \tag{5.56}$$

$$\dot{\lambda}_8 = -q_8x_8 - (1/M_2)\lambda_2 + (2/D_2)\lambda_8 \tag{5.57}$$

$$\dot{\lambda}_9 = -q_9x_9 - r_1k_{I1}(\beta_1k_{P1}x_1 + k_{P1}x_3 + k_{I1}x_9) + (k_{I1}/T_{g1})\lambda_4 \tag{5.58}$$

$$\dot{\lambda}_{10} = -q_{10}x_{10} - r_2k_{I2}(\beta_2k_{P2}x_2 - k_{P2}x_3 + k_{I2}x_{10}) + (k_{I2}/T_{g2})\lambda_5 \tag{5.59}$$

$$\dot{\lambda}_{11} = 0 \tag{5.60}$$

$$\dot{\lambda}_{12} = 0 \tag{5.61}$$

with

$$\lambda_i(t_f) = 0, \qquad i = 1, 2, \ldots, 10 \tag{5.62}$$

$$\lambda_{11}(t_f) = 2w_{11}x_{11}(t_f) \tag{5.63}$$

$$\lambda_{12}(t_f) = 1 \tag{5.64}$$

The components of the gradient vector are given by

$$H_{kP1} = -r_1(\beta_1x_1 + x_3)(k_{P1}\beta_1x_1 + k_{P1}x_3 + k_{I1}x_9) + (1/T_{g1})(\beta_1x_1 + x_3)\lambda_4 \tag{5.65}$$

$$H_{kP2} = -r_2(\beta_2x_2 - x_3)(k_{P2}\beta_2x_2 - k_{P2}x_3 + k_{I2}x_{10}) + (1/T_{g2})(\beta_2x_2 - x_3)\lambda_5 \tag{5.66}$$

$$H_{kI1} = -r_1x_9(k_{P1}\beta_1x_1 + k_{P2}x_3 + k_{I1}x_9) + (1/T_{g1})x_9\lambda_4 \tag{5.67}$$

$$H_{kI2} = -r_2x_{10}(k_{P2}\beta_2x_2 - k_{P2}x_3 + k_{I2}x_{10}) + (1/T_{g2})x_{10}\lambda_5 \tag{5.68}$$

Equations (5.33)-(5.59) completely specify the optimal equations for the LFC of IPS. These equations constitute a two-point boundary value problem. The following algorithm is used to solve this problem, using the method of the conjugate gradient (Refs. 5.11, 5.12).

5.2.4. Solution Technique

Step 1. Select an initial guess for the control parameters,

$$K^0 = \text{col}(k_P^0, k_I^0) \tag{5.69}$$

let the iteration index i be zero.

Step 2. Using the nominal $k^{(i)}$, integrate the state equations (5.18), (5.19), (5.22) from t_0 to t_f with initial conditions $X(t_0) = 0$; store the resulting state trajectory $X^{(i)}$ as a piecewise-constant vector function.

Step 3. Let

$$g_i = \partial H^{(i)}/\partial K^{(i)} \tag{5.70}$$

integrate the costate equations (5.50)-(5.61) backward from t_f to t_0, and evaluate g_i.

Step 4. If

$$\|g_i\| \leq \gamma \tag{5.71}$$

where γ is a preselected positive constant and

$$\|g_i\|^2 = \int_{t_0}^{t_f} g_i^T g_i \, dt \tag{5.72}$$

terminate the iterative procedure, and output the optimal control parameters K. If the stopping criterion is not satisfied, generate a new control variable

$$K^{(i+1)} = K^{(i)} + \alpha_i d_i \tag{5.73}$$

where

$$d_i = g_i + [g_i^T g_i / g_{i-1}^T g_{i-1}] d_{i-1} \tag{5.74}$$

$$d_0 = -g_0 \tag{5.75}$$

Here α_i is determined by solving the one-dimensional minimization problem

$$\min_{\alpha} J(K + \alpha d) \tag{5.76}$$

Equation (5.73) can be solved using Davidon's algorithm (Ref. 5.7). Replace $K^{(i)}$ by $K^{(i+1)}$, and return to Step 2.

5.2.5. Practical Example

The system canonical equations have been programmed for the computer with system parameters (Ref. 5.3)

$$M_1 = 0.04, \qquad G_1 = 0.01, \qquad T_{g1} = 0.5$$
$$T_{t1} = 0.5, \qquad E_1 = 0.03$$
$$M_2 = 0.03, \qquad G_2 = 0.008, \qquad T_{g2} = 1.2$$
$$T_{t2} = 0.5, \qquad D_2 = 0.5, \qquad E_2 = 0.013$$

and performance index weighting coefficients

$$q_i = 1, \qquad i = 1, 2, \ldots, 8$$
$$r_i = 1, \qquad i = 1, 2$$

The penalty factor w_{11} in the case where $x_{11}(t_f)$ is greater than zero has been chosen to be

$$w_{11} = 1000$$

To solve the two-point boundary value problem given by equations (5.33)–(5.68) one can employ the conjugate-gradient algorithm given in the above section.

In the case where area 1 (steam) is subjected to a different step load change and area 2 (hydro) is disturbance-free, it has been found that the optimal control parameters of the steam area (k_{P1}, k_{I1}) are independent of the magnitude and the sign of the area disturbance. For example, by employing the gradient vector as determined by equations (5.65) or (5.67) over $[t_0, t_f] = (0, 20\text{ s})$ for a load disturbance of ± 0.005 p.u., solution by conjugate gradient has been accomplished in five iterations. The initial guesses of the control parameter vector and the step size were taken equal to 0 and 10^{-2}, respectively. The optimal gain parameters have been found to be

$$k_{P1} = 0.337, \qquad k_{I1} = 0.513$$

and the corresponding cost is

$$J = 0.01436$$

The gradient norm is

$$\|H_{K1}\| = 0.2827 \times 10^{-4}$$

with

$$K_1^T = (k_{P1}, k_{I1})$$

The augmented dynamical system eigenvalues are given as follows:

$$+0.01485 \pm j1.82, \qquad -3.303 \pm j0.097, \qquad 2.8668$$
$$-0.311 \pm j0.8418, \qquad 0.641 \pm j0.068$$

Consequently, the optimal PI controller results in a nonasymptotically stable system. However, by obliterating the proportional part and applying the optimization procedure with

$$k_{P1} = 0$$

the optimal integral control parameter obtained has been found to be

$$k_{I1} = 0.5373$$

and the corresponding cost is

$$J = 0.01778$$

The gradient norm is

$$\|H_{kI1}\| = 0.1 \times 10^{-5}$$

The augmented system eigenvalues are given as follows:

$$-0.04996 \pm j1.751, \qquad -0.2194 \pm j7725, \qquad -3.368$$

$$-0.7229 \pm j0.29145, \qquad -3.095, \qquad -2.9016$$

It was found that the optimal control parameter k_{I1} is independent of the load changing. Figures 5.1-5.3 show the system transient and steady state performance in the case of subjecting area 1 to a load change of −0.005 p.u.

In the case where area 2 (hydro) is subjected to variable step load changing and the area 1 (steam) regulators are inoperative, it was also found that the hydroarea optimal control parameters are independent of the load changes. For instance, if area 2 is undergoing a load disturbance of ±0.005 p.u., then by using equations (5.66) and (5.68) over a time interval $[t_0, t_f] = [0, 20\text{ s}]$, the solution by conjugate gradient was completed in four

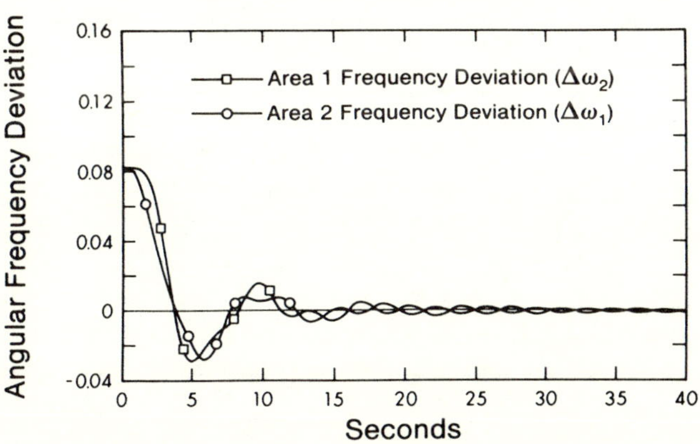

Figure 5.1. Angular frequency deviation versus time.

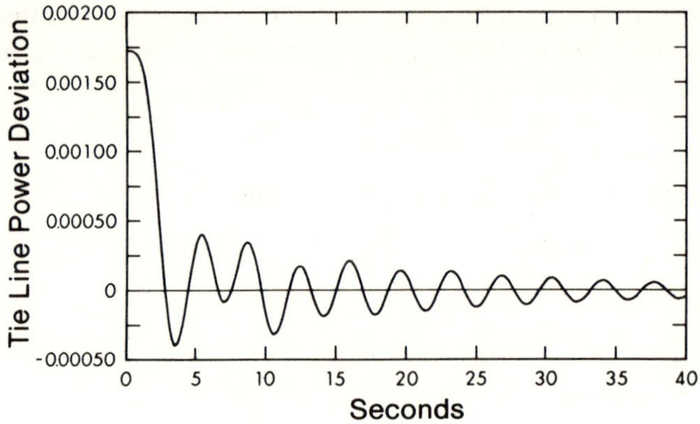

Figure 5.2. Tie-line power deviation versus time.

iterations. The initial guesses of the control parameter vector and the step size were taken equal to 0 and 10^{-2}, respectively. The optimal control parameters of the hydroarea are as follows:

$$k_{P2} = 0.6629, \qquad k_{I2} = 0.348$$

and the corresponding cost is

$$J = 0.0206$$

The gradient norm is

$$\|H_{K2}\| = 0.54 \times 10^{-4}$$

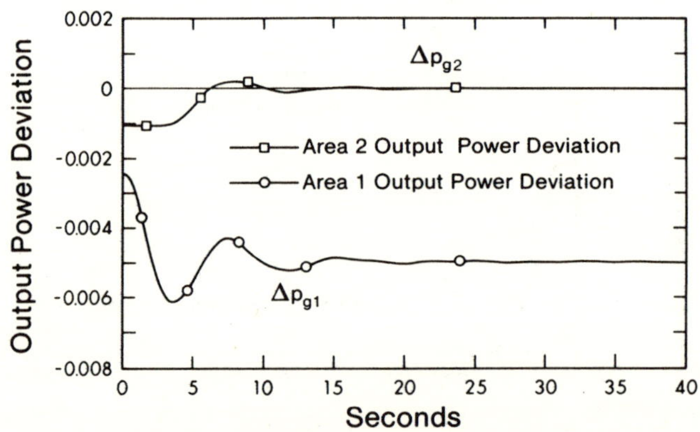

Figure 5.3. Output deviation versus time.

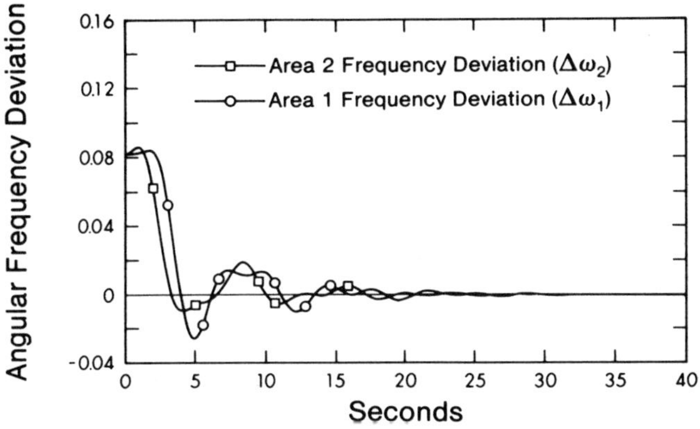

Figure 5.4. Angular frequency deviation versus time.

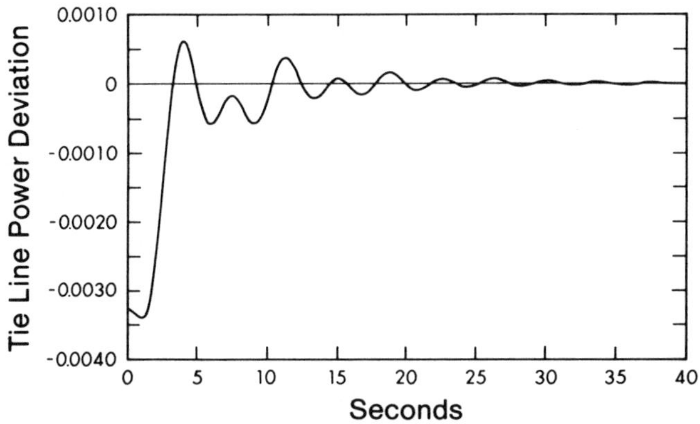

Figure 5.5. Tie-line power deviation versus time.

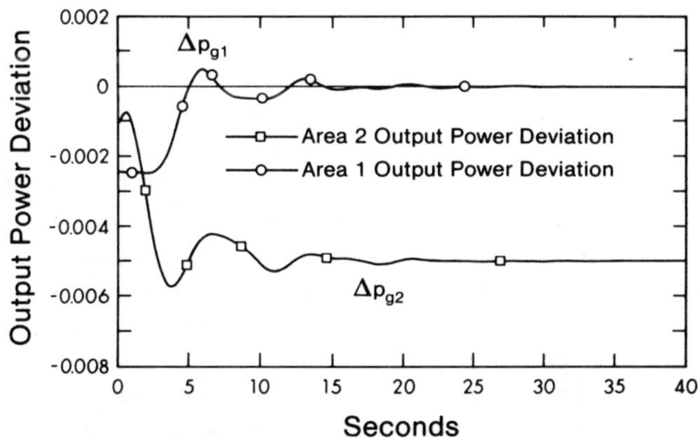

Figure 5.6. Output power deviation versus time.

with

$$K_2^T = (k_{P2}, k_{I2})$$

The augmented system eigenvalues are

$$-3.264 \pm j0.4596, \qquad -2.817, \qquad -1.086$$

$$-0.3058, \qquad -0.2 \pm j.882, \qquad -0.1054 \pm j1.681$$

Figures 5.4–5.6 show the system transient and steady state performance in the case of area 2 being subjected to -0.005 p.u. load changing.

5.3. Optimum Load Frequency Control of Multiarea Interconnected Power Systems

The LFC of an area of an IPS relies on an operating schedule that is usually prepared 24 hr in advance. This schedule indicates the expected demand profile of the area as well as the area commitment to its adjacent areas. The area commitment is the tie-line power interchange which should be maintained at a certain point in time. These values are fed to the area controller as governor set points.

In the normal mode of operation, the area prevailing frequency and tie-line power are compared to the area set point, scheduled frequency, and tie-line power, either every 2–5 s or if the frequency error deviates from a certain band of frequency (±0.033 Hz). In both cases, an area control error (ace), which represents the surplus or the deficiency of the area generation, is computed as follows:

$$\text{ace} = \Delta P_t + \beta\Delta\omega \tag{5.77}$$

where ΔP_t is the area net interchange tie-line power deviation, $\Delta\omega$ is the angular frequency deviation, and β is the area bias-setting. The ace is converted to a train of raise/lower commands which drives the speed changer of the governing system in accordance with the sign of the ace. A negative ace causes generation to increase and vice versa. The speed governor shaft position will change in small steps. At the end of the control period, the speed governing characteristic will be shifted to a new position, a new set point, which results in matching the load demand with the load consumption (Ref. 5.13).

Therefore, the governor set points are altered either at regular intervals or if it is necessary to keep area frequency and tie-line power at their scheduled values. The transition from one set point to another represents a step change in the load demand which itself introduces a frequency transient and tie-line power disturbance before settling down to scheduled values. Therefore, it is necessary to design the area's transients as well as meet the area's steady state specifications. The steady state specifications are met when

- Generation matches the load demand.
- The system frequency deviation is within certain tolerance.
- Interarea flows match a prescribed schedule.

According to the North American Power Systems Interconnection Committee (NAPSIC), ε is equal to 0.033 Hz.

On most modern interconnections, the regulating responsibilities of each area of an interconnection are based on the tie-line-bias concept (conventional LFC) introduced over 25 years ago (Ref. 5.13). When operating in accordance with the principle of conventional control, each area of an IPS attempts to regulate its ace to zero. This can be achieved by employing a PI control of the ace provided that each area can fully accommodate its load changing. The problem of selecting the optimal control parameters of the conventional control strategy was studied in detail by Kirchmayer (Ref. 5.10). In Ref. 5.10, the problem of a two-area IPS was simulated on an analog computer. Then by trial and error, optimal control parameters were found that introduced reasonable maximum overshoot, rise time, settling time, and so on for the frequency and tie-line power deviations.

Previous studies have attempted to control the frequency and the tie-line power deviations independently rather than regulating the ace based on the frequency bias, which is the goal of tie-line-bias control in actual use. References 5.5, 5.6, 5.8, and 5.14 attempt to narrow the gap between conventional and advanced LFC by letting the integral of the ace be a part of the control strategy (Ref. 5.7). In Ref. 5.5, the proportional part was a function of all the system state variables. In Ref. 5.8, both the proportional and integral parts were a function of the output state variables. The necessary conditions to validate the developed control law depend upon the choice of a certain weighting matrix. However, there is no systematic way of computing this matrix. This article treats the automatic generation control (AGC) problem, which basically consists of the LFC problem and the economic dispatch (ED) problem, as an integral one. The ED problem is a "tertiary" control, i.e., it is relatively slow and is based on a quasistatic description of generation and demand (Ref. 5.16). Therefore, the ED should be treated separately from LFC.

In this section the tie-bias-control concept, which is extensively used, is adopted. The control law of each area of a power system is assumed to be PI of the ace. The problem is then formulated using parameter optimization.

5.3.1. Problem Formulation

In order to devise a control scheme for the LFC of an IPS, the basic function of LFC is, of course, regulating the system frequency and the scheduled tie-line power deviation to zero by matching the power system area's generation to the load demand with minimum system transient behavior and control effort. This is a typical optimal control problem. Therefore, the problem of the LFC of an IPS can be expressed mathematically as follows. Given the system dynamics of the IPS,

$$\dot{x} = F(X, U, \Delta L), X(t_0) = 0 \tag{5.78}$$

Here, X is an n-dimensional vector whose components are the dependent variables, U is an m-dimensional control vector, and ΔL is an m-dimensional disturbance vector. It is now necessary to find the optimal control that minimizes the performance index

$$J = \tfrac{1}{2} \int_{t_0}^{t_f} (X^T Q X + U^T R U)\, dt \tag{5.79}$$

The first term in the integrand of equation (5.79) represents the system's transient; the second term represents the control effort. Here, Q and R are positive definite diagonal matrices. LFC exhibits many nonlinearities such as tie-line, governor characteristic, governor deadband, and valves. In this section it is assumed that the IPS is subjected to a small load disturbance and therefore the system dynamics can be linearized around its operating point. That is, the dynamics of the system can be written as follows:

$$\dot{x} = AX + BU + \Gamma \tag{5.80}$$

$$y = CX \tag{5.81}$$

In actual practice, IPS are multiarea systems. Each area is composed of several generators. When an IPS undergoes disturbance, the IPS may be represented by a two-area interconnected system. One area is the disturbed area, connected to the rest of the system (the other area) via a tie-line. Each area may be represented by one generator and its associated prime mover and control devices. Also, each area generator is represented (simulated) with equivalent generator parameters (e.g., equivalent generator inertia is equal to the summation of the inertia of the area generators). Therefore, the control parameters of an area's individual generators are equal to the parameters obtained for the equivalent area generator.

In the case of a TMAIPS, the system parameters are given as follows (Ref. 5.17):

$$A = \begin{bmatrix} -\dfrac{G_1}{M_1} & & -\dfrac{1}{M_1} & & & \dfrac{1}{M_1} & & \\ & -\dfrac{G_2}{M_2} & \dfrac{1}{M_1} & & & & & \dfrac{1}{M_2} \\ T_{12} & -T_{12} & & & & & & \\ -\dfrac{E_1}{T_{g1}} & & & -\dfrac{1}{T_{g1}} & & & & \\ & -\dfrac{E_2}{T_{g2}} & & & -\dfrac{1}{T_{g2}} & & & \\ & & & \dfrac{1}{T_{t1}} & & -\dfrac{1}{T_{t1}} & & \\ & & & & \dfrac{1}{T_{t2}} & & -\dfrac{1}{T_{t2}} & \\ & & & & -\dfrac{2}{D_2} & & \dfrac{2}{T_{t2}}+\dfrac{2}{D_2} & -\dfrac{2}{D_2} \end{bmatrix} \tag{5.82}$$

$$B^T = \begin{bmatrix} 0 & 0 & 0 & -1/T_{g1} & 0 & 0 & 0 & 0 \\ 0 & 0 & 0 & 0 & -1/T_{g2} & 0 & 0 & 0 \end{bmatrix} \tag{5.83}$$

$$\Gamma = \begin{bmatrix} -\Delta L_1/M_1 & 0 & 0 & 0 & 0 & 0 & 0 & 0 \\ 0 & -\Delta L_2/M_2 & 0 & 0 & 0 & 0 & 0 & 0 \end{bmatrix} \tag{5.84}$$

$$C = \begin{bmatrix} 1 & 0 & 0 & 0 & 0 & 0 & 0 & 0 \\ 0 & 1 & 0 & 0 & 0 & 0 & 0 & 0 \\ 0 & 0 & 1 & 0 & 0 & 0 & 0 & 0 \end{bmatrix} \tag{5.85}$$

in which

$$X^T = (x_1, x_2, \ldots, x_8) \tag{5.86}$$

$$U^T = (u_1, u_2) \tag{5.87}$$

where x_1 is the frequency deviation of area 1, x_2 is the frequency deviation of area 2, x_3 is the power angle deviation, x_4 is the deviation in the governor position of area 1, x_5 is the deviation in the governor position of area 2, x_6 is the deviation in mechanical output of area 1, x_7 is the deviation in the gate position of area 2, x_8 is the deviation in mechanical output of area 2, and u_1 is the area 1 speed changer control command and u_2 is the area 2 speed changer control command. The parameters that appear in the preceding matrices are defined as follows: M is the inertial constant in p.u. power s^2/rad, G is the load damping coefficient in p.u. power s/rad, T_1 is the turbine time constant in seconds, T_g is the governor time constant in seconds, D is the water inertia time constant in seconds, T is the synchronizing power in p.u. power/rad, and E is the governor gain in p.u. power s/rad.

The control vector U is specified to be PI of the ace. In terms of the system's state variables, the ace can be written as

$$\text{ace} = DX \tag{5.88}$$

where

$$D = \begin{bmatrix} \beta_1 & 0 & 1 & 0 & 0 & 0 & 0 & 0 \\ 0 & \beta_2 & -1 & 0 & 0 & 0 & 0 & 0 \end{bmatrix}$$

Hence, the control law can be written as follows:

$$U(t) = -K_P DX - K_I \int_{t_0}^{t} DX\, dt \tag{5.89}$$

where

$$K_P = \text{diag}(k_{P1}, k_{P2}) \tag{5.90}$$

$$K_I = \text{diag}(k_{i2}, k_{i2}) \tag{5.91}$$

To cast the system dynamics into the familiar state space form, a new variable vector Z can be defined by

$$Z(t) = \int_{t_0}^{t} DX(t)\, dt \tag{5.92}$$

Hence, the modified system dynamics are given by

$$\dot{X} = (A - BK_P D)X - BK_I Z + \Gamma, \qquad X(t_0) = 0 \tag{5.93}$$

$$\dot{Z} = DX, \qquad Z(t_0) = 0 \tag{5.94}$$

and the modified performance index is given as

$$J = \tfrac{1}{2} \int_{0}^{t_f} [X^T Q_1 X + Z^T Q_2 Z + (K_P DX + K_I Z)^T R (K_P DX + K_I Z)]\, dt \tag{5.95}$$

The second term of J is introduced to penalize the additional state variables Z in the system dynamics.

Now, the optimal control parameters K_P and K_I must be found which minimize the cost functional, equation (5.95), subject to the system dynamic constraints, equations (5.93) and (5.94).

5.3.2. Optimal Solution

To handle a criterion of the form of equation (5.95), the system dynamics are augmented by the additional equation

$$\dot{x}_0 = \tfrac{1}{2}(X^T Q_1 X + Z^T Q_2 Z) + (K_P DX + K_I Z)^T R (K_P DX + K_I Z), \qquad x_0(t_0) = 0 \tag{5.96}$$

By defining a Hamiltonian,

$$H = \sum_{i=1}^{8} \lambda_i x_i + \sum_{j=1}^{2} \lambda_{j+8} \dot{z}_j + \lambda_0 \dot{x}_0 \tag{5.97}$$

Then, by applying Pontryagin's minimum principle, the system canonic equations are obtained. The system state equations are given by equations (5.93) and (5.94). The system costate equations are given as follows:

$$\dot{\lambda}_1 = -q_1 x_1 - r_1 \beta_1 k_{P1}(\beta_1 k_{P1} x_1 + k_{P1} x_3 + k_{I1} z_1) + \frac{G_1}{M_1}\lambda_1 - T_{12}\lambda_3 + \left(\frac{E_1 + \beta_1 k_{P1}}{T_{g1}}\right)\lambda_4 - \beta_1 \lambda_9 \tag{5.98}$$

$$\dot{\lambda}_2 = -q_2 x_2 - r_2 \beta_2 k_{P2}(\beta_2 k_{P2} x_2 - k_{P2} x_3 + k_{I2} z_2) + \frac{G_2}{M_2}\lambda_2 + T_{12}\lambda_3 + \left(\frac{E_2 + \beta_2 k_{P2}}{T_{g2}}\right)\lambda_5 - \beta_2 \lambda_{10} \tag{5.99}$$

$$\dot{\lambda}_3 = -q_3 x_3 - r_1 k_{P1}(\beta_1 k_{P1} x_1 + k_{P1} x_3 + k_{I1} z_1) + r_2 k_{P2}(\beta_2 k_{P2} x_2 - k_{P2} x_3 + k_{I2} z_2) + \frac{1}{M_1}\lambda_1 - \frac{1}{M_2}\lambda_2 + \frac{k_{P1}}{T_{g1}}\lambda_4 - \frac{k_{P2}}{T_{g2}}\lambda_5 - \lambda_9 + \lambda_{10} \tag{5.100}$$

$$\dot{\lambda}_4 = -q_4 x_4 + \frac{1}{T_{g1}}\lambda_4 - \frac{1}{T_{t1}}\lambda_6 \tag{5.101}$$

$$\dot{\lambda}_5 = -q_5 x_5 + \frac{1}{T_{g2}}\lambda_5 - \frac{1}{T_{t2}}\lambda_7 + \frac{2}{T_{t2}}\lambda_8 \tag{5.102}$$

$$\dot{\lambda}_6 = -q_6 x_6 - \frac{1}{M_1}\lambda_1 + \frac{1}{T_{t1}}\lambda_6 \tag{5.103}$$

$$\dot{\lambda}_7 = -q_7 x_7 + \frac{1}{T_{t2}}\lambda_7 - \left(\frac{2}{T_{t2}} + \frac{2}{D_2}\right)\lambda_8 \tag{5.104}$$

$$\dot{\lambda}_8 = -q_8 x_8 - \frac{1}{M_2}\lambda_2 + \frac{2}{D_2}\lambda_8 \tag{5.105}$$

$$\dot{\lambda}_9 = -q_9 x_9 - r_1 k_{I1}(\beta_1 k_{P1} x_1 + k_{P1} x_3 + k_{I1} z_1) + \frac{k_{I1}}{T_{g1}}\lambda_4 \tag{5.106}$$

$$\dot{\lambda}_{10} = -q_{10} x_{10} - r_2 k_{I1}(\beta_2 k_{P2} x_2 - k_{P2} x_3 + k_{I2} z_2) + \frac{k_{I2}}{T_{g2}}\lambda_5 \tag{5.107}$$

$$\dot{\lambda}_0 = 0 \tag{5.108}$$

with

$$\lambda_i(t_f) = 0, \qquad i = 1, 2, \ldots, 10 \tag{5.109}$$

The gradient of the Hamiltonian with respect to the control parameters is given by

$$H_{k_{P1}} = -r_1(\beta_1 x_1 + x_3)(k_{P1}\beta_1 x_1 + k_{P1}x_3 + k_{I1}z_1) + \frac{1}{T_{g1}}(\beta_1 x_1 + x_3)\lambda_4 \tag{5.110}$$

$$H_{k_{P2}} = -r_2(\beta_2 x_2 - x_3)(k_{P2}\beta_2 x_2 - k_{P2}x_3 + k_{I2}z_2) + \frac{1}{T_{g2}}(\beta_2 x_2 - x_3)\lambda_5 \tag{5.111}$$

$$H_{k_{I1}} = -r_1 x_9(k_{P1}\beta_1 x_1 + k_{P2}x_3 + k_{I1}z_1) + \frac{1}{T_{g1}}x_9\lambda_4 \tag{5.112}$$

$$H_{k_{I2}} = -r_2 x_{10}(k_{P2}\beta_2 x_2 - k_{P2}x_3 + k_{I2}z_2) + \frac{1}{T_{g2}}x_{10}\lambda_5 \tag{5.113}$$

Equations (5.98)–(5.113) with equations (5.93) and (5.94) completely specify the optimal control of a TMAIPS. These equations constitute a two-point boundary value problem; the algorithm mentioned in Section 5.2.3 is used again to solve this problem. When finding the optimal control parameters of a TMAIPS, three cases have been considered.

5.3.3. A Sudden Loss or Gain of Load at One Area

The system canonic equations have been programmed for the computer with system parameters (Ref. 5.16)

$$M_1 = 0.04, \quad G_1 = 0.01, \quad T_{g1} = 0.5, \quad T_{t1} = 0.5, \quad E_1 = 0.03$$

$$M_2 = 0.03, \quad G_2 = 0.008, \quad T_{g2} = 1.2, \quad T_{t2} = 0.5, \quad D_2 = 0.5, \quad E_2 = 0.013$$

and

$$q_i = 1, \qquad i = 1, \ldots, 10$$

$$r_i = 1, \qquad i = 1, 2$$

When one area, for example area 1, of the TMAIPS is being subjected to a step load change ($\pm\Delta L_1$), it is assumed that the disturbed area supplementary regulator will take all the regulating responsibilities while the remote area supplementary regulator is not operating. That is,

$$u_1 = k_{P1}\,\mathrm{ace}_1 - k_{I1}\int_{t_0}^{t_f} \mathrm{ace}_1\,dt \tag{5.114}$$

$$u_2 = 0$$

in which

$$\mathrm{ace}_1 = \Delta P_{t12} + \beta_1 \Delta\omega_1$$

Two schemes have been used to design area 1's supplementary regulator:

- Designing an optimal supplementary regulator with frequency bias setting equal to the area natural governing characteristic.
- Designing an optimal supplementary regulator with optimal frequency bias setting.

In the first scheme, the bias setting is given by the following identity:

$$\beta_1 = E_1 + G_1 \tag{5.115}$$

In this case, the control parameter vector is defined by

$$K_1^T = (k_{P1}, k_{I1}) \tag{5.116}$$

The optimal control parameters were found to be independent of the load variations. For example, if area 1 is subjected to a ±0.005 p.u. load change, the conjugate-gradient was completed in four iterations over a time interval $[t_0, t_f]$ equal to 0.20 s (at $t_f = 20$ the system reaches the steady state). The initial guesses of the control parameter vector and the step size were taken equal to 0 and 10^{-2}, respectively. The optimal control parameters are as follows:

$$K_1^T = (0.08335, 0.451) \tag{5.117}$$

and the corresponding cost

$$J = 0.0242$$

with a gradient norm

$$H_{k_1} = 0.2 \times 10^{-4}$$

In the second scheme, the problem of designing an optimal supplementary regulator optimum frequency bias setting is considered. This is reduced to a single-parameter optimization problem. The control gain parameters

k_{P1} and k_{I1} are taken from identity equation (5.118). In this case, the gradient is given as

$$H_{\beta_1} = -r_1 k_{P1} x_1[k_{P1}(\beta_1 x_1 + x_3) + k_{I1} x_9] + \frac{1}{T_{g1}} x_2 k_{P1} \lambda_4 - x_1 \lambda_9 \quad (5.118)$$

Adopting the technique shown in Section 5.2.3, the optimal bias setting is found to be independent of the load changing. In the case of a ±0.005 p.u. load change, the initial guess of the frequency bias setting and the initial step size equal 0.04 (natural governing characteristic) and 0.1, respectively. After three iterations, the optimal bias setting is given as

$$\beta_1 = 0.03595$$

and the corresponding cost

$$J = 0.02417$$

with a gradient norm

$$H_{\beta_1} = 0.3133 \times 10^{-5}$$

Therefore, the value of the optimum frequency bias setting (0.036) is less than the value of the area natural governing characteristic (0.04). The effect of the area bias setting on the value of the cost functional, the optimality criterion, is negligible. Similarly, the problem of finding the optimal control parameters in the case that area 2 is being subjected to load change ($\pm\Delta L_2$) and the remote area supplementary regulator is not operating results in

$$\begin{aligned} u_1 &= 0 \\ u_2 &= -k_{P2}\,\mathrm{ace}_2 - k_{I2}\int_{t_0}^{t_f} \mathrm{ace}_2\, dt \end{aligned} \quad (5.119)$$

in which

$$K_2^T = (k_{P2}, k_{I2}) \quad (5.120)$$

For example, if area 2 is being subjected to a load change of ±0.005 p.u., the corresponding cost is given by

$$J = 0.0302$$

with a gradient norm

$$H_{k_2} = 0.38639 \times 10^{-4} \quad (5.121)$$

The initial control parameter and the initial step size were taken to be 0 and 10^{-1}, respectively. When designing the area 2 controller with optimum bias setting, the gradient of the Hamiltonian with respect to the bias setting is given by

$$H_{\beta_2} = -r_2 k_{P2} x_2 [k_{P2}(\beta_2 x_2 - x_3) + k_{I2} x_{10}] + \frac{1}{T_{g2}} x_2 k_{P2} \lambda_5 - x_2 \lambda_{10} \quad (5.122)$$

Using the gradient as determined in equation (5.122) with the control gain parameters given as

$$K_2^T = (0.607, 0.32)$$

over

$$[t_0, t_f] = [0, 20 \text{ s}]$$

the conjugate-gradient descent was accomplished with the program described in Section 5.2.3. The initial guess of the frequency bias setting is taken equal to twice the natural governing characteristic (0.021). After seven iterations, the optimal frequency bias setting is

$$\beta_2 = 0.0148$$

with a corresponding cost corresponding to a ±0.005 p.u. load disturbance

$$J = 0.029985$$

and a gradient norm

$$\|H_{\beta_2}\| = 0.141 \times 10^{15}$$

Once more, it can be seen that the effect of employing a tie-line bias controller with optimum frequency bias setting produces negligible effect in the system optimality criterion, J. For example, in the case of a ±0.005 p.u. load change, the loss of optimality, β,

$$\beta \doteq [(J_{\text{opt.}\,\beta_2} - J_{\text{nonopt.}\,\beta_2})/J_{\text{nonopt.}\,\beta_2}] \times 100 \quad (5.123)$$

is equal to 0.712%.

Figures 5.7–5.9 show the system performance when area 1 is subjected to a −0.005 p.u. load change and the area 2 supplementary regulator is not operating.

5.3.4. A Loss of a Tie-Line between Two Areas

This is equivalent to a sudden load increase in one area and a sudden load decrease in the other. The system parameters before and after the removal of the tie-line are the same when the line is dc and not the same when the line is ac. In the latter case, both areas' supplementary regulators

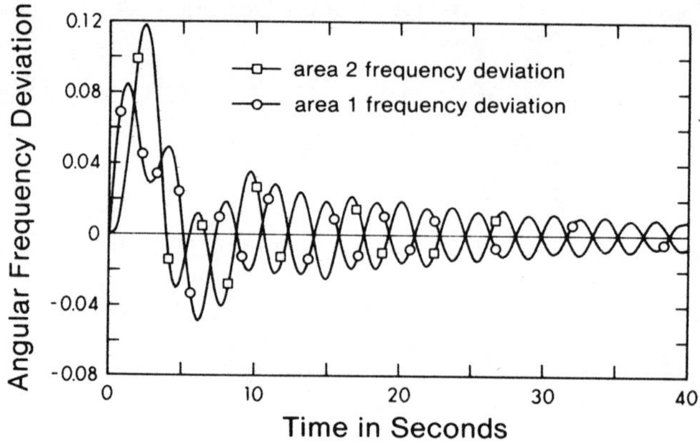

Figure 5.7. Frequency deviation versus time.

should come into action simultaneously. Accordingly, the control law of the two-area IPS can be written as follows:

$$\begin{aligned} u_1 &= -k_{P1}\,\mathrm{ace}_1 - k_{I1}\int_{t_0}^{t_f} \mathrm{ace}_1\,dt \\ u_2 &= -k_{P2}\,\mathrm{ace}_2 - k_{I2}\int_{t_0}^{t_f} \mathrm{ace}_2\,dt \end{aligned} \tag{5.124}$$

For example, if the system is subjected to a load disturbance,

$$\Delta L_1 = -0.005$$

$$\Delta L_2 = +0.005$$

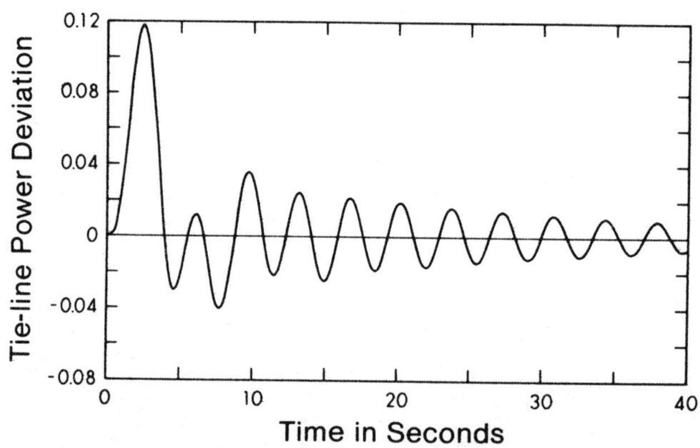

Figure 5.8. Tie-line power deviation versus time.

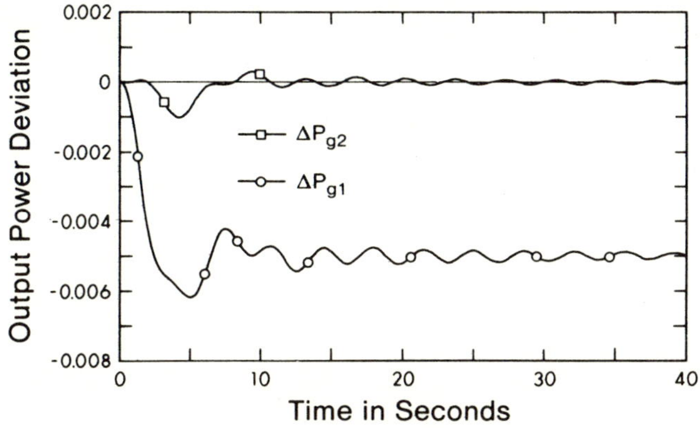

Figure 5.9. Output power deviation versus time. System dynamic response of a TMAIPS obtained from interactive governor-supplementary regulator: $\Delta L_1 = -0.005$ p.u., $u_1 = -0.0834\,\text{ace}_1 - 0.451 \int_{t_0}^{t_f} \text{ace}_2\, dt$, $\Delta L_2 = 0.0$, $u_2 = 0$.

which may represent a sustained removal of one of the dc tie-lines connecting area 1 to area 2.

The solution of the canonic equations given by equations (5.93), (5.94), and (5.98)-(5.113) subject to the boundary conditions

$$x_i(t_0) = 0, \qquad i = 1, 2, \ldots, 10$$

$$\lambda_i(t_f) = 0, \qquad i = 1, 2, \ldots, 10, \lambda_0(t_f) = 1$$

yields the control parameter vector

$$K = \text{col}(k_{P1}, k_{P2}, k_{I1}, k_{I2}) \tag{5.125}$$

By using the technique described in Section 5.2.3 the optimum value of the control parameter vector is found to be

$$K = \text{col}(-0.4484, -0.0025, 0.3472, 0.442)$$

and the value of the optimum cost function

$$J = 0.0192$$

with a gradient norm

$$\|H_K\| = 0.971 \times 10^{-4}$$

The initial guesses of the control parameter and the initial step size have been taken equal to 0 and 0.1, respectively. In the case of employing an integral control action, that is,

$$K = \text{col}(k_{I1}, k_{I2})$$

the optimal control parameters are

$$K = \mathrm{col}(0.255, 0.27)$$

and the value of the optimum cost function is

$$J = 0.0276$$

when applying the control parameters obtained in Ref. 5.17, which are given as follows:

$$k_{I1} = 0.09$$

$$k_{I2} = 0.4$$

the corresponding cost is

$$J = 0.06946$$

which is almost 2.5 times that of the proposed controller. Figures 5.10 and 5.11 show a comparison between the system performances in both cases.

5.3.5. A Sudden Loss of Generation at One Area

This is equivalent to the previous situation except that the system parameters before and after the disturbance are not the same. The value of the disturbed area inertia constant is lower after the occurrence of the disturbance. For instance, if area 1 undergoes loss of $-\Delta L_1$ p.u. of its generation, the new value of M_1 is

$$M_1^n = M_1(1 - \Delta L_1) \tag{5.126}$$

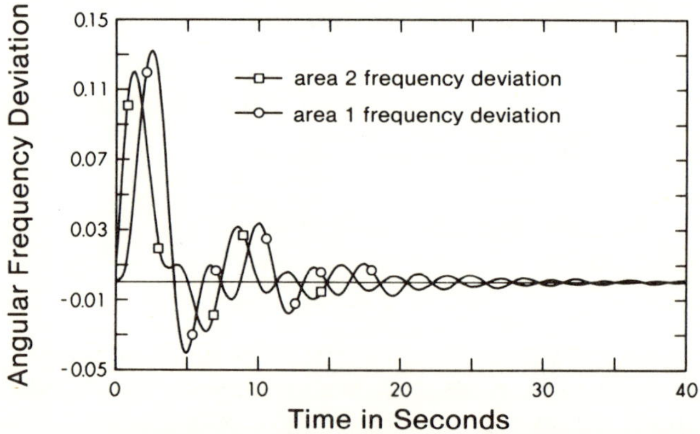

Figure 5.10. Frequency deviation versus time.

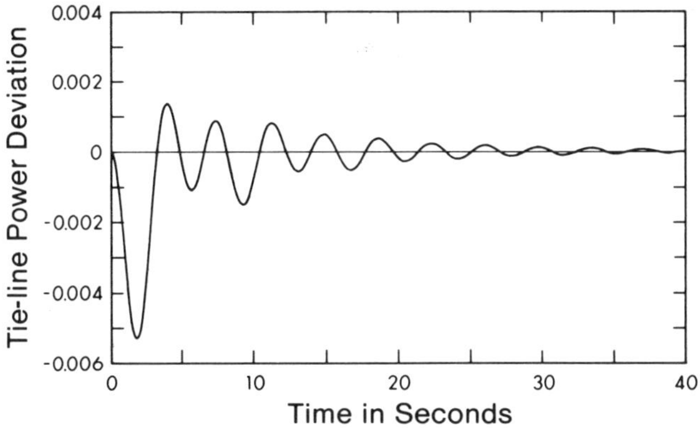

Figure 5.11. Tie-line power deviation.

in which M_1^n is the area 1 inertial constant after the occurrence of the disturbance, and M_1 is the area 1 inertia constant before the occurrence of the disturbance.

The optimal control parameters depend upon the values of the areas' inertial constants. Table 5.1 shows the relationship between area 1 inertia constant and the optimal control parameters. Table 5.2 shows the same for area 2. However, the optimal control parameters that were obtained when the areas' inertia is constant result in a stable system with a slight loss of optimality compared to the case where the areas' inertia is considered variable. For example, if area 1 is being subjected to a 0.1 p.u. loss of its generation and area 1 is employing a supplementary regulator with gain parameters

$$k_{P1} = 0.0833$$

$$k_{I1} = 0.451$$

Table 5.1. Steam Area Optimum Control Parameters versus the Area Inertia

Generation loss	Inertia constant	k_{P1}	k_{I1}
−0.005	0.0398	0.0833	0.4510
−0.01	0.0396	0.0823	0.4511
−0.02	0.0392	0.0792	0.4513
−0.04	0.0384	0.0755	0.4514
−0.06	0.0376	0.0721	0.4512
−0.08	0.0368	0.0694	0.4506
−0.1	0.036	0.0673	0.450

Table 5.2. Hydroarea Optimum Control Parameters versus the Area Inertia

Generation loss	Inertia constant	k_{P2}	k_{I2}
−0.005	0.02985	0.605	0.321
−0.01	0.0297	0.603	0.321
−0.02	0.0294	0.500	0.322
−0.04	0.0288	0.592	0.324
−0.06	0.0282	0.585	0.326
−0.08	0.0276	0.57	0.328
−0.10	0.027	0.57	0.33

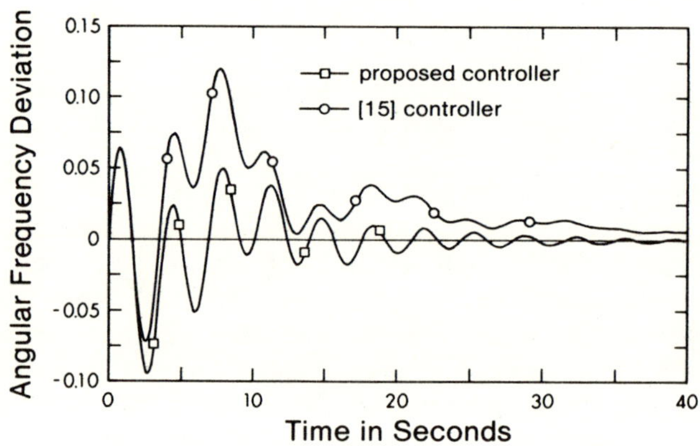

Figure 5.12. Area 1 frequency deviation versus time.

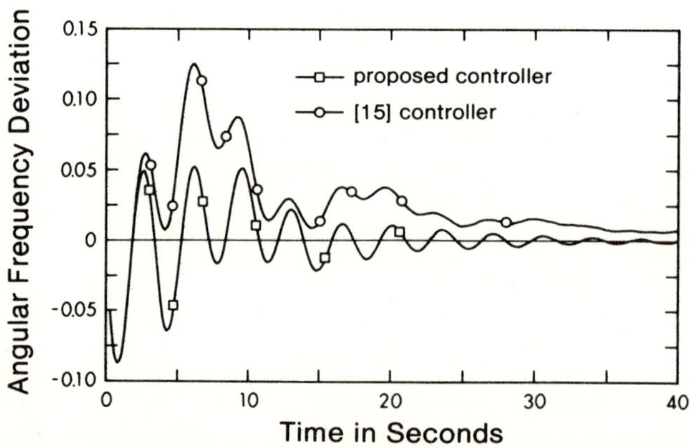

Figure 5.13. Area 2 frequency deviation versus time.

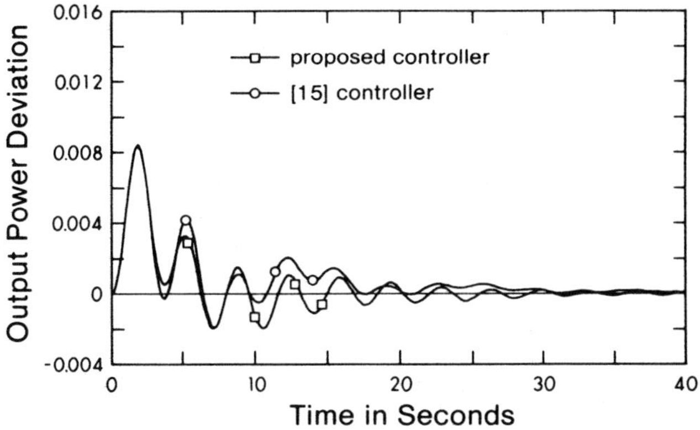

Figure 5.14. Tie-line power deviation versus time.

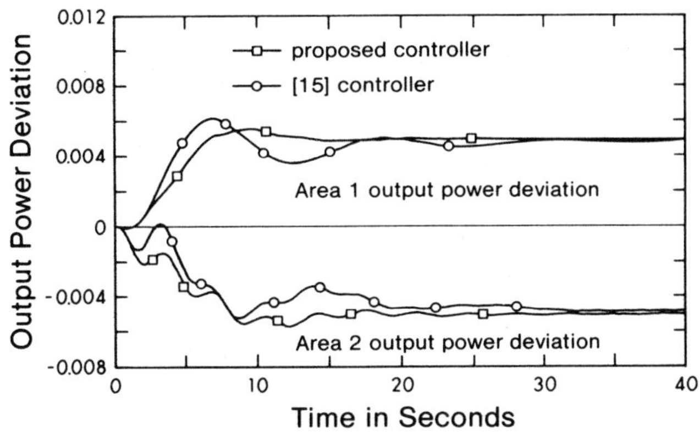

Figure 5.15. Output power deviation versus time.

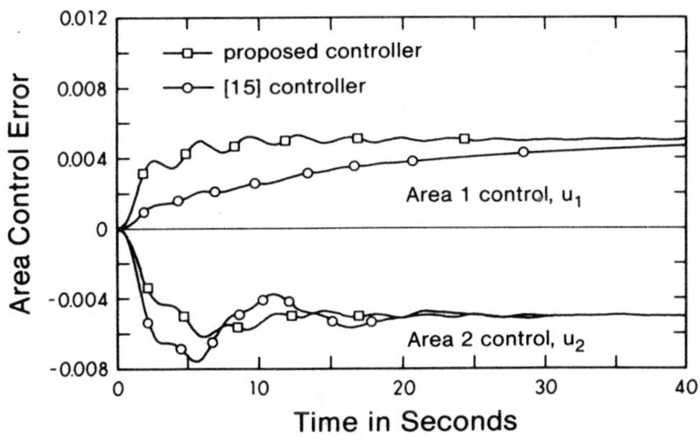

Figure 5.16. Area control error versus time.

which resulted from solving the system canonic equations with the assumption that the system parameters before and after the disturbance are the same, that is

$$M_1^n = M_1 = 0.04$$

this results in the cost function

$$J_1 = 0.099361 \times 10^{+1}$$

when it is applied to the system with $M_1^n = 0.036$. When the system parameters before and after are not considered the same,

$$M_1 = 0.04$$

$$M_1^n = 0.036$$

The optimal control parameters are given as

$$k_{P1} = 0.0673$$

$$k_{I1} = 0.45$$

and the value of the optimal cost function is

$$J_2 = 0.9928 \times 10^{+1}$$

Hence, when employing a control strategy that neglects the changes in the

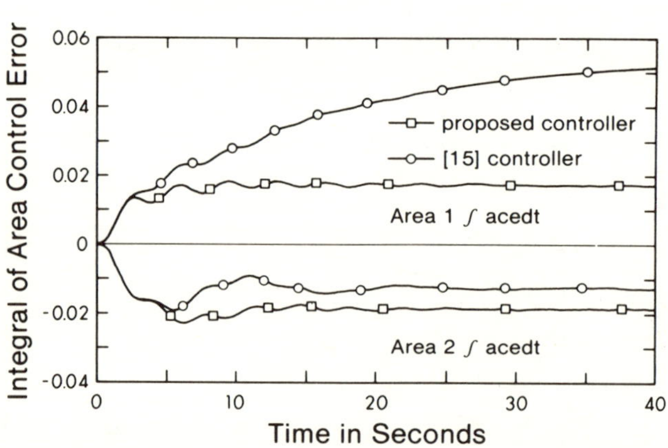

Figure 5.17. $\int$ ace dt versus time. System dynamic response of a TMAIPS obtained from the proposed controller: $u_1 = -0.255 \int \text{ace}_1\, dt$; $u_2 = 0.27 \int \text{ace}_2\, dt$; and the controller advised in Ref. 5.15: $u_1 = -0.09 \int \text{ace}_1\, dt$, $u_2 = -0.4 \int \text{ace}_2\, dt$, $\Delta L_1 = -0.05$ p.u., $\Delta L_2 = 0.005$ p.u.

system's inertia, optimality is lost:

$$\beta = \frac{J_1 - J_2}{J_1} \times 100 = 0.0815\%$$

Therefore, the assumption that the system's inertia is equal before and after a generation disturbance is satisfactory (see Figs. 5.12–5.17).

5.4. Optimal Load Frequency Control with Governor Backlash

A well-developed linear control theory exists, resulting in more research being directed toward including nonlinear aspects of general control systems. Linear control theory (classical or optimal) suffers from the fundamental criticism that, in reality, dynamical systems are frequently subject to several complicating factors that may invalidate or at least severely limit its applicability. There are no general methods for analysis and synthesis of nonlinear control systems in classical or modern control theory. For example, to employ modern control theory when dealing with systems with control signal saturation (relay control systems, Refs. 5.18 and 5.19) is quite different from employing modern control theory when dealing with systems with output feedback (for example), which contains a functional nonlinear element (Ref. 5.20).

Here, we deal with a system whose dynamical equations comprise a hysteresis element. The approach is general and will be applied to the problem of LFC of a single steam area power system.

A single-area system is one in which load changes are accommodated by the system as a whole, regardless of where they occur on the system. Load changes that occur in any part of the system may be absorbed elsewhere within the system, in accordance with allocation practices prevailing at that particular time. No one part of the system is expected to adjust its own generation to counteract its own load changes.

Most large power systems are geographically widely spread, made up of different operating companies. Experience in the automatic operation of these systems sometimes shows a desirable need to assign the regulation of the system frequency to one of the companies in the group centrally located with respect to the other companies. Regulation of the remaining companies in the system depends on the frequency and the power of the tie lines connecting the respective companies to the central area. Therefore, the central area tie-line power flows are neither scheduled nor controlled. LFC which adopts complete relaxation of the tie-line power interchange between the companies of a system is referred to as flat frequency control (FFC).

5.4.1. Mathematical Model

A typical model of a single steam area power system with governor deadband is shown in Fig. 5.18 (Ref. 5.10). The governor deadband is the hysteresis effect which arises from mechanical friction and/or looseness in the moving parts of the speed-governing system. That is, for any given controlled valve position, there may be a change in the control signal direction before the valve position changes.

In state variable form, the system dynamical equations can be written as

$$\dot{X} = f(X, u, \Delta L, t), \qquad X(t_0) = 0 \tag{5.127}$$

$$Y = CX \tag{5.128}$$

where

$$X^T = (x_1, x_2, x_3) \tag{5.129}$$

$$C = (1, 0, 0) \tag{5.130}$$

Here, the symbols are as follows: $x_1 = \Delta\omega$ is the angular frequency deviation in rad/s; $x_2 = \Delta x_g$ is the deviation in governor position in p.u. power; $x_3 = \Delta P_g$ is the deviation in turbine output power in p.u. power; u is the speed changer position in p.u. power; ΔL is the step load change.

The control is specified to be integral or PI of the angular frequency deviation. In terms of the system state variables, the control law can be written as follows:

$$u = -k_P x_1 - k_I \int_{t_0}^{t_f} x_1\, dt \tag{5.131}$$

Here, the symbols are as follows: k_P is the proportional gain; k_I is the integral gain. The control parameters k_P and k_I can be expressed mathematically as follows:

$$\dot{K} = 0 \tag{5.132}$$

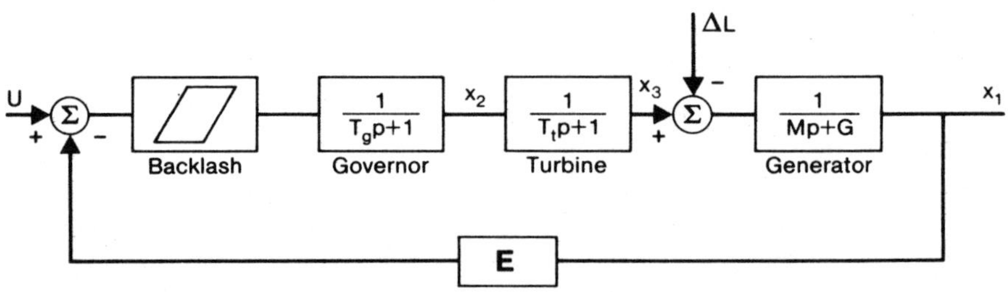

Figure 5.18. Block diagram of a single steam area with backlash.

where

$$K^T = (k_P, k_I) \tag{5.133}$$

To cast the system dynamical equations into the familiar state space form, one can introduce a new variable,

$$x_4 = \int_{t_0}^{t_f} x_1 \, dt \tag{5.134}$$

Therefore, u can be written as

$$u = k_P x_1 - k_I x_4 \tag{5.135}$$

Now, we include the backlash element explicitly in the system dynamics. For notational convenience, new variables y_i, $i = 1, 2, 3, 4, 5$, are introduced, such that

$$y_1 = x_1, \qquad y_2 = u - Ex_1, \qquad y_3 = x_2, \qquad y_4 = x_3, \qquad y_5 = x_4 \tag{5.136}$$

Therefore, the system dynamics can be written as

$$\dot{y}_1 = -(G/M)y_1 + (1/M)y_4 - \Delta L/M \tag{5.137}$$

$$\begin{aligned}\dot{y}_2 = &-[(k_P + E)(G/M) + k_I]y_1 \\ &- (k_P/M)y_4 + k_P(\Delta L/M) + E(\Delta L/M)\end{aligned} \tag{5.138}$$

$$\dot{y}_3 = (1/T_g)y_{20} - (1/T_g)y_3 \tag{5.139}$$

$$\dot{y}_4 = (1/T_t)y_3 - (1/T_t)y_4 \tag{5.140}$$

$$\dot{y}_5 = y_1 \tag{5.141}$$

in which

$$y_{20} = \begin{cases} y_2 - \Delta, & \dot{y}_2 < 0 \\ y_2 + \Delta, & \dot{y}_2 > 0 \\ \text{const}, & |y_{20} - y_2| \leq \Delta \end{cases} \tag{5.142}$$

The problem posed is to find the optimal control parameters k_P and k_I, which minimize the cost functional

$$J = (1/2) \int_{t_0}^{t_f} \left[\sum_{i=1}^{5} q_i y_i^2 + r(k_P y_1 + k_I y_5)^2 \right] dt \tag{5.143}$$

subject to the dynamical constraints given by equations (5.137)–(5.142).

5.4.2. Problem Formulation

To handle a criterion of the form (5.143) a new state variable is defined,

$$y_6 = (1/2)\left\{ \sum_{i=1}^{5} q_i y_i^2 + r(k_P y_1 + k_I y_5)^2 \right\} \tag{5.144}$$

with

$$y_6(t_0) = 0 \tag{5.145}$$

Therefore, the cost functional can be written as follows:

$$J = y_6(t_f) \tag{5.146}$$

Now, we define a Hamiltonian

$$H = \sum_{i=1}^{6} \lambda_i y_i \tag{5.147}$$

Then, we apply Pontryagin's minimum principle. The system state equations are given by equations (5.137)-(5.142), (5.144), and (5.145). The system costate equations are found to be

$$\begin{aligned} \dot{\lambda}_1 &= -q_1 y_1 + -rk_P(k_P y_1 + k_I y_5)(G/M)\lambda_1 \\ &\quad - [(k_P + E)(G/M) + k_I]\lambda_2 \end{aligned} \tag{5.148}$$

$$\dot{\lambda}_2 = -q_2 x_2 - S(1/T_g)\lambda_3 \tag{5.149}$$

$$\dot{\lambda}_3 = -q_3 y_3 + (1/T_g)\lambda_3 - (1/T_t)\lambda_4 \tag{4.150}$$

$$\dot{\lambda}_4 = -q_4 y_4 - (1/M)\lambda_1 + (k_P/M)\lambda_2 + (1/T_t)\lambda_4 \tag{4.151}$$

$$\dot{\lambda}_5 = -rk_I(k_P y_1 + k_I y_5) \tag{4.152}$$

$$\dot{\lambda}_6 = 0 \tag{4.153}$$

with

$$\begin{aligned} \dot{\lambda}_i(t_f) &= 0, \qquad i = 1, 2, \ldots, 5 \\ \dot{\lambda}_6(t_f) &= 1 \end{aligned} \tag{4.154}$$

in which

$$S = \begin{cases} 0, & \text{deadband is active} \\ 1, & \text{otherwise} \end{cases} \tag{5.155}$$

The gradient vector components are given by

$$H_{k_P} = -ry_1(k_P y_1 + k_I y_5) - [(G/M)y_1 - \Delta L/M]\lambda_2 \tag{5.156}$$

$$H_{k_I} = -ry_5(k_P y_1 + k_I y_5) - y_1\lambda_2 \tag{5.157}$$

Since the integrand in the cost functional equation (5.143) and the system dynamical equations are not explicit functions of t, we have

$$\dot{H} = 0 \tag{5.158}$$

or

$$H = \text{const} \tag{5.159}$$

on the optimal trajectory. Therefore, in the deadband zone, the following relations should be satisfied:

$$[\partial H/\partial y_i]_{t_d^-} = [\partial H/\partial y_i]_{t_d^+}, \qquad i = 1, 2, \ldots, 6 \tag{5.160}$$

that is

$$\lambda_i(t_d^-) = \lambda_i(t_d^+), \qquad i = 1, 2, \ldots, 6 \tag{5.161}$$

and

$$H_{k_P}(t_d^-) = H_{k_P}(t_d^+) \tag{5.162}$$

In equations (5.161) and (5.162)

$$H_{k_I}(t_d^-) = H_{k_I}(t_d^+) \tag{5.163}$$

$$t_d = \lim_{\varepsilon \to 0}(t_d - \varepsilon) \tag{5.164}$$

$$t_d^+ = \lim_{\varepsilon \to 0}(t_d + \varepsilon) \tag{5.165}$$

with

$$\dot{y}_2(t_d^-) = y_2(t_d^+) = 0 \tag{5.166}$$

the symbol t_d denotes the time at which the deadband starts or ceases to be active. Equations (5.161) and (5.162) are the Weierstrass-Erdmann corner conditions. A corner point of $y(t)$ is a point at which the derivative dy/dt poses a jump discontinuity.

Equations (5.137)-(5.142) with equations (5.148)-(5.166) completely specify the optimal solution. These equations constitute a two-point boundary value problem with multiple corner points. The following technique is used to solve this problem.

5.4.3. Solution Technique

The problem of finding the optimal control parameters for the PI controller of a dynamical system which includes a backlash element has been reduced to the problem of solving a nonlinear, two-point boundary value problem with multiple corner points.

To solve a two-point boundary value problem with multiple corner points, the technique in Section 5.2.3 will be adopted with a slight modification. This modification will take place in the forward integration to facilitate the detection of the corner points. As the integrating process proceeds in the forward direction, the values of $y_2(t_i)$ and $\dot{y}_2(t_j)$, $j = i - 2$, $i - 1$, are stored. When a new point $y_2(t_i)$ is being obtained, two problems have to be solved before proceeding with the integration process. These are (a) detection of the mode changing, which may be classified as follows:

(i) rising mode of operation, (ii) falling mode of operation, (iii) deadband mode of operation; and (b) locating the exact times of mode transition, i.e., the times of entering and leaving the deadband (corner points).

Detection of the change of the mode of operation can be easily done as follows:

(i) In the case of entering the deadband mode, the sign of the prevailing $\dot{y}_2$ is opposite to the previous one; that is,

$$\text{sign}\{\dot{y}_2(t_i)\dot{y}_2(t_{i-1})] = -ve$$

(ii) In the case of leaving the deadband mode, the following inequality should be satisfied:

$$|y_2(t_r) = y_2(t_i)| \geq \Delta$$

in which $y_2(t_r)$ = last reversing point.

(iii) In the case of a rising mode of operation,

$$\text{sign}\{\dot{y}_2(t_i)\} = +ve$$

(iv) In the case of a falling mode of operation

$$\text{sign}\{\dot{y}_2(t_i)\} = -ve$$

The problem of finding the exact time of entering the deadband, reversing $\dot{y}_2(t_r)$, can be done by employing some interpolation technique. Assume that, at the present point t_i, the relation

$$\text{sign}\{\dot{y}_2(t_i)\dot{y}_2(t_{i-1})\} = -ve$$

is satisfied. This means that the system has already passed the reversing point t_r. To determine t_r, one can employ a Hermite third-order interpolation polynomial to approximate $y_2(t)$ through $y_2(t_{i-2})$ and $y_2(t_{i-1})$. The third-order Hermite polynomial is given by the following equation:

$$\hat{y}_2(t) = \sum_{j=i-2}^{i=1} y_2(t_j)H^*_{i-1,j}(t) + \sum_{j=i-2}^{i-1} \dot{y}_2(t_j)\hat{H}_{i-1,j}(t) \tag{5.167}$$

in which

$$H^*_{r-1,j} = [1 - 2(t - t_j)L'_{i-1,j}(t_j)]L^2_{r-1,j}(t) \tag{5.168}$$

$$\hat{H}_{r-1,j} = (t - t_j)L^2_{r-1,j}(t) \tag{5.169}$$

$$L_{n,k} = \prod_{\substack{j=0\\ j\neq k}}^{i=1} [(t - t_j)/(t_k - t_j)] \tag{5.170}$$

$$L' = dL/dt \tag{5.171}$$

This yields the equation

$$\hat{x}_2(t) = at^3 + bt^2 + ct + d \tag{5.172}$$

in which

$$a = \{-2[y_2(t_i - 1) - y_2(t_{i-2})] + (t_{i-1} - t_{i-2}) \times [\dot{y}_2(t_{i-1}) + \dot{y}_2(t_{i-2})]\}/(t_{i-1} - t_{i-2})^3 \tag{5.173}$$

$$b = \{3(t_{i-1} + t_{i-2})[y_2(t_{i-1}) - y_2(t_{i-2}) + (t_{i-1} - t_{i-2})] \times [(t_{i-1} + 2t_{i-2})\dot{y}_2(t_{i-1}) + 2(t_{i-1} + t_{i-2})\dot{y}_2(t_{i-2})]\}/(t_{i-1} - t_{i-2})^3 \tag{5.174}$$

$$c = \{-Gt_{i-1}t_{i-2}[y_2(t_{i-1}) - y_2(t_{i-2})] + (t_{i-1} - t_{i-2})[(2t_{i-1}t_{i-2} + t_{i-1}^2) \times \dot{y}_2(t_{i-1}) + (t_{i-1}^2 + 2t_{i-1})\dot{y}_2(t_{i-1})]\}/(t_{i-1} - t_{i-2})^3 \tag{5.175}$$

$$d = \{(3t_{i-1} - t_{i-2})t_{i-2}^2 y_2(t_{i-1}) + (t_{i-1} - 3t_{i-2})t_{i-1}^2 y_2(t_{i-2}) - (t_{i-1} - t_{i-2}) \times t_{i-1}t_{i-2}[t_{i-2}\dot{y}_2(t_{i-1}) + t_{i-1}\dot{y}_2(t_{i-2})]\}/(t_{i-1} - t_{i-2})^3 \tag{5.176}$$

The reversing point t_r occurs when

$$d\hat{y}_2(t)/dt = 3at_r^2 + 2bt_r + c = 0$$

or

$$t_r = -b/3a + \sqrt{(b^2/9a^2 - c/3a)} \tag{5.177}$$

Knowing the reversing time, the program reintegrates to obtain $y_2(t_r)$. If

$$|y_2(t_r) - \hat{y}_2| \leq \varepsilon \tag{5.178}$$

where ε is a predetermined limit, then $\hat{y}_2(t_r)$ or $y_2(t_r)$ is accepted as a corner point. If not, $y_2(t_{i-1})$ will be replaced by $y_2(t_r)$, and the process is repeated until the reversing point is located accurately. Then $y_2(t_r)$ is stored.

The problem of finding the exact time of leaving or entering the deadband can be solved by employing an inverse interpolation process. Once again, the Hermite third-order polynomial will be used iteratively to find the exact point of entering or leaving the deadband. This can be done by solving the following equation:

$$y_2(t_r) \pm 2\Delta = at_d^3 + bt_d^2 + ct_d + d \tag{5.179}$$

where t_d is the time at which the deadband starts to be inactive. Equation (5.179) can be solved analytically, but it has been found that the analytical solution accuracy is dependent on the coefficients a, b, c.

The Newton–Raphson method has been employed to solve equations (5.179) instead of the analytical method. As in the case of finding reversing points, the process of finding t_d will be iteratively repeated until t_d is determined accurately.

Having solved the problems of finding the reversing, entering, and leaving points, one can solve the two-point boundary value problem given by equations (5.137)–(5.142) and equations (5.148)–(5.166) using the conjugate-gradient algorithm, discussed in Section 5.2.3, to get the optimal control parameters.

5.4.4. Practical Example

The system was subjected to different load disturbances. It was found that the effect of the backlash element in the system dynamics on the optimal control gain parameters is overriding for small load disturbances. In the case where the system is being subjected to a load disturbance of magnitude somewhat greater than 0.05 p.u. for the following parameters:

$$T_g = 0.5, \qquad T_t = 0.5, \qquad M = 0.04, \qquad G = 0.01, \qquad E = 0.03$$

$$q_i = 1, \qquad i = 1, \ldots, 4, \qquad r = 1$$

it was found that the optimal control gain parameters with and without taking into consideration the backlash effect are the same. For load disturbances of magnitude somewhat less than 0.05 p.u., the optimal control gain parameters depend on the disturbance magnitude. For example, if the system is being subjected to a load disturbance of −0.004 p.u., the optimal control parameters, when the backlash element was taken into account in the system dynamics, were found to be

$$k_P = 0.0567, \qquad k_I = 0.0297$$

and the cost functional was found to be

$$J_0 = 0.639 \times 10^{-2}$$

In the case of neglecting the backlash nonlinearity, the optimal control parameters are found to be independent of the load variations. These optimal control parameters are given as

$$k_P = 0.086, \qquad k_I = 0.022$$

When these linear optimal control parameters are employed to control the nonlinear system, the cost functional was found to be

$$J_s = 0.725 \times 10^{-2}$$

Therefore, by adopting a suboptimal strategy, there is a loss in the optimality performance index as follows:

$$\varepsilon = [(J_s - J_0)/J_0] \times 100\% = 13.45\%$$

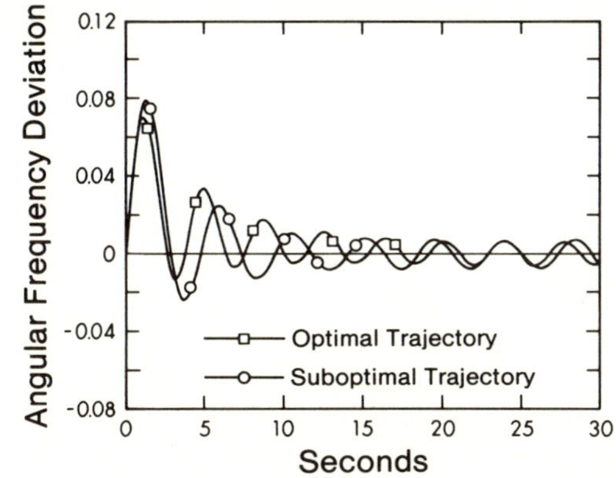

Figure 5.19. Angular frequency deviation versus time.

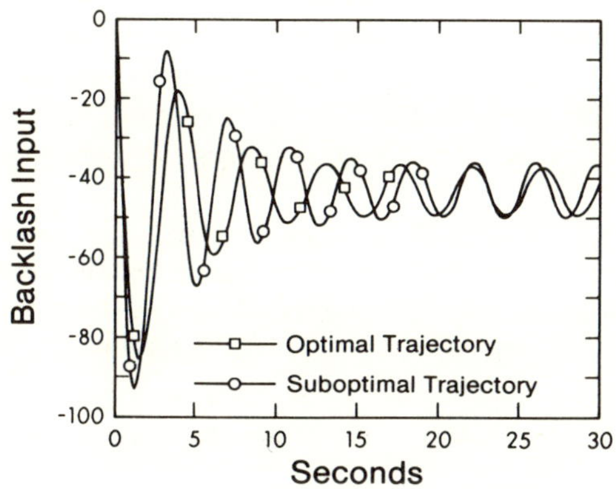

Figure 5.20. Backlash input versus time.

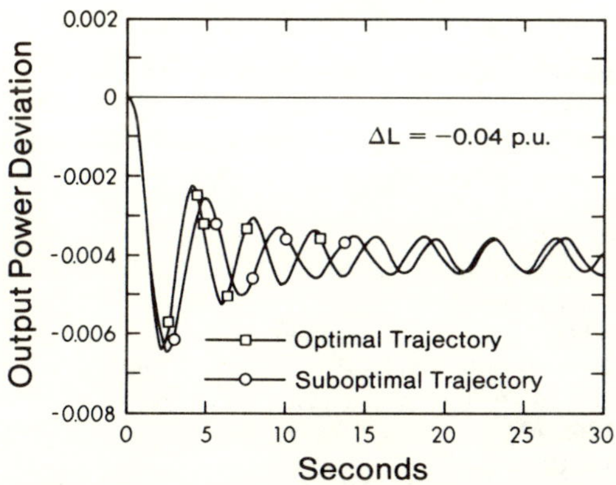

Figure 5.21. Output power deviation versus time.

A comparison between the system behaviors in the two cases is shown in Figs. 5.19–5.21.

5.5. Conclusions

In this chapter, the solution of the optimal tie-line control problem was discussed. The problem of the load frequency control of interconnected power systems with a prespecified control of proportional-plus-integral action was discussed and reduced in a systematic way to a two-point boundary-value problem. The solution of this problem leads to the values of the optimal control parameters.

The problem of designing an optimal controller for solving the load frequency control of a two-area interconnected power system has been studied for different load disturbances. The control law is specified to be of proportional and/or integral action (present-day practice). The problem of finding the optimal control has been reduced to solving a two-point boundary value problem. It has been shown that, for each kind of load disturbance, the optimal control parameters are independent of the load variations.

The problem of finding the optimal control for a system that incorporates a backlash nonlinearity can be handled by employing Pontryagin's minimum principle and the Weierstrass–Erdmann corner conditions. The problem of detecting the corner points was solved by employing interpolation and extrapolation techniques. The remaining part of the problem is to solve the two-point boundary value problem. This was done by employing the conjugate-gradient technique.

References

5.1. Elgerd, O. I., and Fosha, C. E., "Optimum Megawatt Frequency Control of Multi-Area Electric Energy Systems," *IEEE Trans. Power Appar. Syst.* **PAS-89**, 556–563 (1970).

5.2. Fosha, C. E., and Elgerd, O. I., "The Megawatt-Frequency Control Problem: A New Approach via Optimal Control Theory," *IEEE Trans. Power Appar. Syst.* **PAS-89**, 563–571 (1970).

5.3. Miniesy, S. M., and Bohn, E. V., "Optimum Load Frequency Continuous Control with Unknown Deterministic Power Demand," *IEEE Trans. Power Appar. Syst.* **PAS-91**, 1910–1915 (1972).

5.4. Newmann, M. M., "Optimal and Suboptimal Control Using an Observer When Some of the State Variables Are Not Measurable," *Int. J. Control* **9**, 281–290 (1969).

5.5. Calovic, M. S., "Power System Load Frequency Control Using an Optimum Linear Regulator with Integral Feedback," Proceedings of the 5th World IFAC Congress, Toulouse, France, 1972.

5.6. Calovic, M. S., "Linear Regulator Design for Load Frequency Control," *IEEE Trans. Power Appar. Syst.* **PAS-91**, 2271–2285 (1972).

5.7. Porter, B., "Optimal Control for Multivariable Linear Systems Incorporating Integral Feedback," *Electron. Lett.* **6**, 815–816 (1970).

5.8. Calovic, M. S., Cuk, N. M., and Djorovic, M., "Autonomous Area Generation Control of Interconnected Power Systems," *Proc. IEEE* **124**, 392–402 (1977).

5.9. Ewart, D. N., "Automatic Generation Control, Performance under Normal Conditions," Proceedings of the Engineering Foundation Conference, Henniker, New Hampshire, 1971.

5.10. Kirchmayer, L. K., *Economic Control of Interconnected Systems*, Wiley, New York, 1959.

5.11. Kirk, D. E., *Optimal Control Theory: An Introduction*, Prentice-Hall, Englewood Cliffs, New Jersey, pp. 154–161, 1970.

5.12. Fox, R. L., *Optimization Methods for Engineering Design*, Addison-Wesley, Reading, Massachusetts, 1971.

5.13. Cohn, N., *Control of Generation and Power Flow on Interconnected Power Systems*, Wiley, New York, 1971.

5.14. Calovic, M. S., "An Application of Optimal Linear Regulators with Prescribed Steady State," *Int. J. Control* **1972**, 801–816.

5.15. Cohn, N., "Some Aspects of Tie-Line Bias Control Interconnected Power System," *Trans. AIEE* **1957**, 1415–1436.

5.16. Quazza, G., "Automatic Control in Electric Power Systems," *Automatica* **6**, 123–150 (1970).

5.17. Aggarwall, R. P., and Bergserth, F. R., "Large Signal Dynamics of LFC and their Optimization Using Non-Linear Programming I," *IEEE Trans. Power Appar. Syst.* **PAS-87**, 527–532 (1968).

5.18. Flugge-Lotz, I., *Discontinuous and Optimal Control*, McGraw-Hill, New York, 1968.

5.19. Ryan, E. P., *Optimal Relay and Saturating Control Synthesis*, IEEE, Control Engineering Series, 1982.

5.20. Anderson, B. O., and Moore, J. B., *Linear Optimal Control*, Prentice-Hall, Englewood Cliffs, New Jersey, 1971.

5.21. Davidon, W. C., "Variable Metric Method for Minimization," Argonne National Laboratory, Report No. ANL-5990 (Revised), 1959.

5.22. Abdel-Halim, M. A., Christensen, G. S., and Kelly, D. H., "Optimal Load Frequency Control with Governor Backlash," *J. Opt. Theory Appl.* **45**(4), (1985).

5.23. Abdel-Halim, M. A., Christensen, G. S., and Kelly, D. H., "Decentralized Optimum Load Frequency Control of Interconnected Power Systems," *J. Opt. Theory Appl.* **45**(4), 517–531 (1985).

5.24. Abdel-Halim, M. A., Christensen, G. S., and Kelly, D. H., "Optimum Load Frequency Control of Multi-Area Interconnected Power Systems," *Can. Elect. Eng. J.* **10**(1), 32–39 (1985).

6

Optimal Control of Turboalternators

6.1. Introduction

This chapter is devoted to the solution of the problem of optimal control of synchronous turboalternators. In the first section of this chapter the optimal torque and voltage control of a large turbogenerator connected to an infinite bus is found by using the minimum norm formulation. In the second section the optimal control of a system with two identical interconnected turbogenerators connected to an infinite bus is considered. Control of the generators is effected through control of field voltages and turbine torques. Finally, in the last section a realistic feedback control of two interconnected turbogenerators is considered.

6.2. Optimal Control of a Large Turboalternator

It is desired to find the optimal torque and voltage control of a synchronous turboalternator connected to an infinite bus. The controls are considered to be applied directly to the machine, i.e., no time constants are considered for either the governor or voltage regulator.

The machine is operating in the steady state and is suddenly subjected to a three-phase short circuit. The machine is then immediately disconnected from the infinite bus for a certain period and is subsequently reconnected assuming that the fault has been cleared. Thus, generally, the machine will experience a disturbance that might even lead to loss of synchronism. It is desired to find optimally the controls needed as functions of time which will minimize a cost functional, quadratic in the various state and control variables.

6.2.1. Mathematical Model

Let the symbol δ denote the torque angle of the machine. Thus if one defines $\dot{\delta} = \eta$, the machine may be modeled by the following system of equations (see Appendix 6.A):

$$\dot{\delta} = \eta \tag{6.1}$$

$$\dot{\eta} = u_1/M - (s_4/M)\sin\delta\cos\delta - (s_5/M)x_3\sin\delta - (k_d/M)\eta \tag{6.2}$$

$$\dot{x}_3 = u_2 - Ax_3 + c\cos\delta \tag{6.3}$$

The assumptions used in the derivation of this third-order system are now summarized: (i) damper windings are not considered—instead, a damping coefficient is introduced in the torque equation; (ii) the transient response in the transmission line is neglected; (iii) the resistances in the stator winding, the transformer, and the transmission line are neglected; (iv) the time derivative of the fluxes ψ_d and ψ_q is neglected—that is, $p\psi_d = 0$, $p\psi_4 = 0$; and (v) the angular velocity ω is approximated by the synchronous velocity ω_0 in the voltage equations.

Equations (6.1)-(6.3) represent a synchronous machine to a very good approximation (Refs. 6.1 and 6.2).

The cost functional to be minimized is

$$J_1 = \int_{t_0}^{t_f} \{\alpha_1(\delta - \delta^s)^2 + \alpha_2\eta^2 + \alpha_3(x_3 - x_3^s)^2 + \beta_1(u_1 - u_1^s)^2 + \beta_2(u_2 - u_2^s)^2\}\, dt \tag{6.4}$$

where the superscript s indicates the steady-state value. The parameters α_1, α_2, α_3, β_1, β_2 are penalty factors for deviation from steady-state values. Because the necessary and sufficient conditions for optimality are easily implemented if the system can be cast into the minimum norm formulation, it is then required to cast equations (6.1)-(6.3) into the quadratic form, because the cost functional is already quadratic. This can be accomplished by expanding $\sin\delta$ and $\cos\delta$ in Taylor series and introducing pseudovariables (Ref. 6.3).

A more general approach, however, is to make a nonlinear transformation on the variable δ. It should be noted that this transformation is applicable to all power system problems, since they contain nonlinearities of the form appearing in equations (6.1)-(6.3).

The transformation introduced is

$$x_1 = \sin\delta \tag{6.5}$$

let

$$x_2 = \eta \tag{6.6}$$

then, equations (6.1)-(6.3) become

$$\dot{x}_1 = (1 - x_1^2)^{1/2} x_2 \tag{6.7}$$

$$\dot{x}_2 = u_1/M - (s_4/M)x_1(1 - x_1^2)^{1/2} - (s_5/M)x_3x_1 - (k_d/M)x_2 \tag{6.8}$$

$$\dot{x}_3 = u_2 - Ax_3 + c(1 - x_1^2)^{1/2} \tag{6.9}$$

To cast the system completely into a quadratic form, the following pseudovariable is introduced:

$$x_4^2 = 1 - x_1^2 \tag{6.10}$$

Then, the equations (6.7)-(6.9) become

$$\dot{x}_1 = x_2 x_4 \tag{6.11}$$

$$\dot{x}_2 = u_1/M - (s_4/M)x_1x_4 - (s_5/M)x_3x_1 - (k_d/M)x_2 \tag{6.12}$$

$$\dot{x}_3 = u_2 - Ax_3 + cx_4 \tag{6.13}$$

$$x_4^2 = 1 - x_1^2 \tag{6.14}$$

Owing to the transformations (6.5) and (6.6), one should express $(\delta - \delta^s)^2$ in the cost functional in terms of $x_1 - x_1^s$ and then expand the resulting function and introduce additional pseudovariables. On the other hand, one could alter the cost functional slightly and, if necessary, increase the factor α, to compensate for this change, as will be shown below.

A cost functional with $(\delta - \delta^s)^2$ replaced by

$$(x_1 - x_1^s)^2 + (x_4 - x_4^s)^2$$

is monotonically increasing in the range $(0, \pi)$ and approximates $(\delta - \delta^s)^2$ to a high degree of accuracy over the range $(p, \pi/2)$. This can be seen as follows:

$$(x_1 - x_1^s)^2 + (x_4 - x_4^s)^2 = 2[1 - \cos(\delta - \delta^s)] \tag{6.15}$$

If one now chooses α, $\alpha = 2.5\alpha_1$, for the coefficient of

$$(x_1 - x_1^s)^2 + (x_4 - x_4^s)^2$$

in the cost functional, one will have the same penalty at π for both the old and the new cost functional. It should be noticed that, in this case, small errors are penalized more heavily than before, which is a beneficial feature of the modified cost functional.

Now, the modified cost functional is

$$J_2 = \int_{t_0}^{t_f} \{\alpha[(x_1 - x_1^s)^2 + (x_4 - x_4^s)^2] + \alpha_2 x_2^s + \alpha_3(x_3 - x_3^s)^2 + \beta_1(u_1 - u_1^s)^2 + \beta_2(u_2 - u_2^s)^2\}\, dt \tag{6.16}$$

The system is now in a quadratic form, and the optimization proceeds by adjoining the equations to the cost functional via Lagrange multipliers. For this particular problem, however, the equality constraint on the pseudovariable x_4 is converted to a differential equation and one will still have a quadratic structure without introducing any additional pseudovariables. Even though the system is increased to fourth order, an advantage is gained in that the equality constraints are unified in structure, so that it is easier to find a unified approach for the numerical solution of the system. To accomplish this, equation (6.14) is differentiated, and one uses equation (6.11). This results in

$$\dot{x}_4 = -x_2 x_1 \tag{6.17}$$

The optimization problem can now be formulated. Given the equations

$$\dot{x}_1 = x_2 x_4 \tag{6.18}$$

$$\dot{x}_2 = (1/M)(u_1 - s_4 x_1 x_4 - s_5 x_1 x_3 - k_d x_2) \tag{6.19}$$

$$\dot{x}_3 = u_2 - A x_3 + c x_4 \tag{6.20}$$

$$\dot{x}_4 = x_2 x_1 \tag{6.21}$$

it is desired to minimize the cost functional

$$J_2 = \int_{t_0}^{t_f} \{\alpha[(x_1 - x_1^s)^2 + (x_4 - x_4^s)^2] + \alpha_2 x_2^2 + \alpha_3 (x_3 - x_3^s)^2 + \beta_1 (u_1 - u_1^s)^2 + \beta_2 (u_2 - u_2^s)^2\}\, dt \tag{6.22}$$

where $x_i(t_0)$ are given. One also has the following limits on the control variables:

$$\begin{aligned} &\text{if } u_1 < 0 \qquad \text{then we put } u_1 = 0 \\ &\text{if } u_1 > 1 \qquad \text{then we put } u_1 = 1 \\ &\text{if } u_2 < 0 \qquad \text{then we put } u_2 = 0 \\ &\text{if } u_2 > 1 \qquad \text{then we put } u_2 = 1 \end{aligned} \tag{6.23}$$

These are inequality constraints and will be incorporated in the system via Kuhn–Tucker multipliers (Ref. 6.4).

6.2.2. A Minimum Norm Formulation

The augmented cost functional is obtained by adjoining equations (6.18)–(6.21) to the cost functional via Lagrange multipliers and the inequality constraints (6.23) via Kuhn–Tucker multipliers. One thus obtains

$$J_3 = \int_{t_0}^{t_f} \{\alpha[(x_1 - x_1^s)^2 + (x_4 - x_4^s)^2] + \alpha_2 x_2^2 + \alpha_3(x_3 - x_3^s)^2$$

$$+ \beta_1(u_1 - u_1^s)^2 + \beta_2(u_2 - u_2^s)^2 + \lambda_1(\dot{x}_1 - x_2x_4)$$

$$+ \lambda_2[\dot{x}_2 - (1/M)(u_1 - s_4x_1x_4 - s_5x_1x_3 - k_dx_2)]$$

$$+ \lambda_3(\dot{x}_3 - u_2 + Ax_3 - cx_4) + \lambda_4(\dot{x}_4 + x_1x_2)$$

$$+ e_1(u_1 - u_{1\max}) + e_1^1(u_{1\min} - u_1) + e_2(u_2 - u_{2\max})$$

$$+ e_2^1(u_{2\min} - u_2)\}\, dt \tag{6.24}$$

Integrating (6.24) by parts and dropping terms that do not depend on $x_i(t)$, $u_i(t)$, and $x_i(t_f)$ [since $x_i(t_f)$ are not specified], one obtains

$$J = \int_{t_0}^{t_f} \{\alpha x_1^2 + \alpha_2 x_2^2 + \alpha_3 x_3^2 + \alpha x_4^2 + \beta_1 u_1^2 + \beta_2 u_2^2$$

$$+ \lambda_2(s_4/M)x_2x_4 + \lambda_2(s_5/M)x_1x_3 - \lambda_1x_2x_4 + \lambda_4x_1x_2$$

$$+ x_1(-2\alpha x_1^s - \dot{\lambda}_1) + x_2(-\dot{\lambda}_2 + \lambda_2(k_d/M) + x_3(-2\alpha_3x_3^s - \dot{\lambda}_3 + A\lambda_3)$$

$$+ x_4(-2\alpha x_4^s - c\lambda_3 - \dot{\lambda}_4) + u_1(-2\beta_1u_1^s - \lambda_2/M + e_1 - e_1^1)$$

$$+ u_2(-2\beta_2u_2^s - \lambda_3 + e_2 - e_2^1)\}\, dt$$

$$+ \{\lambda_1(t_f)x_1(t_f) + \lambda_2(t_f)x_2(t_f) + \lambda_3(t_f)x_3(t_f) + \lambda_4(t_f)x_4(t_f)\} \tag{6.25}$$

It will be noticed that J is composed of an integral part and a boundary part, which are independent of each other.

If one defines the vector X such that

$$X^T = (x_1, x_2, x_3, x_4, u_1, u_2) \tag{6.26}$$

then J can be written as

$$J = \int_{t_0}^{t_f} (X^TLX + R^TX)\, dt + [\lambda_1(t_f)x_1(t_f) + \lambda_2(t_f)x_2(t_f) + \lambda_3(t_f)x_3(t_f)$$

$$+ \lambda_4(t_f)x_4(t_f)] \tag{6.27}$$

where

$$L = \begin{bmatrix} \alpha & \lambda_4/2 & (s_5/2M)\lambda_2 & (s_4/2M)\lambda_2 & 0 & 0 \\ \lambda_4/2 & \alpha_2 & 0 & -\lambda_1/2 & 0 & 0 \\ (s_5/2M)\lambda_2 & 0 & \alpha_3 & 0 & 0 & 0 \\ (s_4/M)\lambda_2 & -\lambda_1/2 & 0 & \alpha & 0 & 0 \\ 0 & 0 & 0 & 0 & \beta_1 & 0 \\ 0 & 0 & 0 & 0 & 0 & \beta_2 \end{bmatrix} \tag{6.28}$$

and

$$\begin{aligned} R^T = [&-2\alpha x_1^s - \dot{\lambda}_1, -\dot{\lambda}_2 + \lambda_2 k_d/M, -\dot{\lambda}_3 + A\lambda_3 + A\lambda_3 - 2\alpha_3 x_3^s, \\ &- \dot{\lambda}_4 - c\lambda_3 - 2\alpha x_4^s, -\lambda_2/M - 2\beta_1 u_1^s + e_1 - e_1^1, \\ &-\lambda_3 - 2\beta_2 u_2^s + (e_2 - e_2^1)] \end{aligned} \tag{6.29}$$

To minimize J in equation (6.27) one minimizes the integral and the boundary parts separately. Hence, one can write

$$\min_{[X, x_i(t_f)]} J = \min_X \int_{t_0}^{t_f} (X^T L X + R^T X)\, dt + \sum_{i=1}^{4} \min_{x_i(t_f)} \lambda_i(t_f) x_i(t_f) \tag{6.30}$$

Now, define

$$V = L^{-1} R \tag{6.31}$$

then, the integral part of equation (18) can be written in the following form, by a process similar to completing the squares:

$$\min_X \hat{J} = \min_X \int_{t_0}^{t_f} \{[X + (V/2)]^T L [X + (V/2)] - V^T L V/4\}\, dt \tag{6.32}$$

Since it is desired to minimize $\hat{J}$ with respect to X, the problem is equivalent to

$$\min_X \hat{J} = \min_X \int_{t_0}^{t_f} \{[X + (V/2)]^T L (X + V/2)\}\, dt \tag{6.33}$$

because V is independent of X. Equation (6.33) defines a norm, hence, one can write equation (6.33) as

$$\min_X \hat{J} = \min_X [\|X + \tfrac{1}{2}V\|]_L \tag{6.34}$$

6.2.3. The Optimal Solution

The minimum of $\hat{J}$ is clearly achieved for

$$[X + \tfrac{1}{2}V] = [0] \tag{6.35}$$

provided the matrix L is positive definite. Thus from equation (6.31), one finds that the optimal solution is given by

$$[R + 2LX] = [0] \tag{6.36}$$

The boundary term in equation (6.27) is minimized when

$$\lambda_i(t_f) = 0 \tag{6.37}$$

because $x_i(t_f)\lambda_i(t_f)$ has a minimum when

$$\delta(x_i(t_f))\lambda_i(t_f) = 0 \tag{6.38}$$

and $\delta(x_i(t_f))$ is arbitrary.

Writing (6.36) explicitly, rearranging, and adding the machine equations, one obtains the following set:

$$\dot{\lambda}_1 = 2\alpha(x_1 - x_1^s) + \lambda_4 x_2 + (s_5/M)\lambda_2 x_3 + (s_4/M)\lambda_2 x_4 \tag{6.39}$$

$$\dot{\lambda}_2 = 2\alpha_2 x_2 - \lambda_1 x_4 + \lambda_4 x_1 + (k_d/M)\lambda_2 \tag{6.40}$$

$$\dot{\lambda}_3 = 2\alpha_3(x_3 - x_3^s) + (s_5/M)\lambda_2 x_1 + A\lambda_3 \tag{6.41}$$

$$\dot{\lambda}_4 = 2\alpha(x_4 - x_4^s) + (s_4/M)\lambda_2 x_1 - \lambda_1 x_2 - c\lambda_3 \tag{6.42}$$

$$\dot{x}_1 = x_2 x_4 \tag{6.43}$$

$$\dot{x}_2 = (1/M)[u_1 - s_4 x_1 x_4 - s_5 x_1 x_3 - k_d x_2] \tag{6.44}$$

$$\dot{x}_3 = u_2 - Ax_3 + cx_4 \tag{6.45}$$

$$\dot{x}_4 = -x_1 x_2 \tag{6.46}$$

$$u_1 = \lambda_2/2\beta_1 M + u_1^s - (e_1^1)/2\beta_1 \tag{6.47}$$

$$u_2 = \lambda_3/2\beta_2 + u_2^s - (e_2 - e_2^1)/2\beta_2 \tag{6.48}$$

One also has the exclusion equations

$$\begin{aligned} e_1(u_1 - u_{1\max}) &= 0, \qquad e_1^1(u_{1\min} - u_1) = 0 \\ e_2(u_2 - u_{2\max}) &= 0, \qquad e_2^1(u_{2\min} - u_2) = 0 \end{aligned} \tag{6.49}$$

where

$$u_{1\max} = u_{2\max} = 1.0, \qquad u_{1\min} = u_{2\min} = 0 \tag{6.50}$$

Equations (6.39)-(6.50) with equation (6.37) completely specify the optimal equations. These equations constitute a two-point boundary value problem. The following section describes the technique used to solve this problem.

6.2.4. Solution Technique

Equations (6.39)-(6.50) are solved by the following method. The two-point boundary value problem is solved on the computer as an initial-value problem by devising a way to estimate closely the unknown initial value of the Lagrange multipliers $\lambda_i(t)$. One may use iterative techniques, such as the steepest descent method or the Newton–Raphson method, to find $\lambda_i(t_0)$. However, it turns out that $\lambda_i(t)$ are very sensitive to the initial values chosen, $\lambda_i(t_0)$; therefore, it is very difficult to match the boundary conditions at $t = t_f$ unless the $\lambda_i(t_0)$ are in very close proximity to the actual values.

To get an estimate for $\lambda_i(t_0)$, it is assumed that the control minimizes the cost functional over (t_0, t_f), so that Δx_i and $\Delta\lambda_i$ are small. Hence, one linearizes the system around the steady-state values and obtains the following system from equations (6.30)–(6.50):

$$\Delta\dot{\lambda}_1 = 2\alpha\,\Delta x_1 + [(s_5/M)x_3^s + (s_4/M)x_4^s]\,\Delta\lambda_2 \tag{6.51}$$

$$\Delta\dot{\lambda}_2 = 2\alpha_2\,\Delta x_2 - x_4^s\,\Delta\lambda_1 + x_1^s\,\Delta\lambda_4 + (k_d/M)\,\Delta\lambda_2 \tag{6.52}$$

$$\Delta\dot{\lambda}_3 = 2\alpha_3\,\Delta x_3 + (s_5/M)x_1^s\,\Delta\lambda_2 + A\,\Delta\lambda_3 \tag{6.53}$$

$$\Delta\dot{\lambda}_4 = 2\alpha\,\Delta x_4 + (s_4/M)x_1^s\,\Delta\lambda_2 - c\,\Delta\lambda_3 \tag{6.54}$$

$$\Delta\dot{x}_1 = x_4^s\,\Delta x_2 \tag{6.55}$$

$$\begin{aligned}\Delta\dot{x}_2 = {} & \Delta\lambda_2/2\beta_1 M^2 - [(s_4/M)x_4^s + (s_5/M)x_3^s]\,\Delta x_1 \\ & - (s_5/M)x_1^s\,\Delta x_3 - (s_4/M)x_1^s\,\Delta x_4 \\ & - (k_d/M)\Delta x_2(1/M)(u_1^s - s_4 x_1^s x_4^s - s_5 x_1^x x_3^s)\end{aligned} \tag{6.56}$$

$$\Delta\dot{x}_3 = \Delta\lambda_3/2\beta_2 - A\,\Delta x_3 + c\,\Delta x_4 + u_2^s - Ax_3^s + cx_4^s \tag{6.57}$$

$$\Delta\dot{x}_4 = -x_1^s\Delta x_2 \tag{6.58}$$

with the boundary conditions

$$\Delta x_i(t_0) = x_i(t_0) - x_i^s, \qquad \Delta\lambda_i(t_f) = 0 \tag{6.59}$$

The objective is now to find $\lambda_i(t_0)$, which will be used as an initial estimate for the nonlinear system of equations (6.39)–(6.50). This can be done via the transition matrix approach; however, as it turns out, the transition matrix has very large and very small numbers which are both important in the determination of $\lambda_i(t_0)$, and a large amount of error would be introduced by using this method because equations (6.51)–(6.59) are ill conditioned. On the other hand, the full transition matrix is not needed, since $\lambda_i(t)$ can be expressed as linear functions of $x_i(t)$ via the Riccati equation, and the problem associated with the transition matrix does not occur here, because the backward integration of the Riccati equation ensures simultaneously the satisfaction of the boundary value problem. Furthermore, the error associated with the transition matrix approach is reduced considerably by this latter method.

6.2.5. Computational Example

The solution techniques mentioned in the above section have been used to find the optimal torque and optimal voltage for a turbogenerator. Steady-state values and machine constants are given in Appendix 6.B. The machine is operating at steady state and is suddenly subjected to a three-phase short circuit. The machine is then immediately disconnected from the infinite bus for a certain time (0.4 s) and is subsequently reconnected.

The problem is optimized over two separate intervals. For the interval from $t = 0$ to $t = 0.4$ s, the machine is disconnected, and the following trial values were obtained for $\lambda_i(0)$ by using the Riccati equation:

$$\lambda_1(0) = -0.0107, \qquad \lambda_2(0) = -0.0665$$

$$\lambda_3(0) = -0.00781, \qquad \lambda_4(0) = 0.00813$$

These values resulted in $\lambda_i(t_f)$ being essentially equal to zero; and, integrating equations (6.39)–(6.50), one thus obtains the solution in Table 6.1.

When the machine was reconnected, it was found that an optimizing interval of 0.605 s was sufficient to bring the system to the steady state. Hence, the interval considered was from $t = 0.4$ to $t = 1.05$ s. The following trial values for $\lambda_i(0.4)$ were obtained by using the Riccati equation:

$$\lambda_1(0.4) = -0.255, \qquad \lambda_2(0.4) = -0.0606$$

$$\lambda_3(0.4) = -0.0906, \qquad \lambda_4(0.4) = 0.161$$

A direct search for the optimal solution around the above values resulted in $\lambda_i(t_f)$ being essentially zero; and integrating equations (6.39)–(6.50) one thus obtains the solution in Table 6.2.

6.3. Optimal Control of Two Interconnected Turbogenerators

In this section a system with two identical interconnected turbogenerators that are connected to an infinite bus (Fig. 6.1) is considered. Control of the turboalternators is effected through control of field voltages and turbine torques. The controls are considered to be applied directly to

Table 6.1. Optimal Solution for Open Switch

t (s)	x_1	x_2	x_3	x_4	u_1	u_2	λ_1	λ_2	λ_3	λ_4
0	0.603	0.0	1.874	0.798	0.0137	0.7341	−0.0108	−0.0664	−0.0078	0.0081
0.5	0.6032	0.0097	1.8947	0.7978	0.0046	0.7341	−0.0107	−0.0672	−0.0078	0.0081
0.1	0.6037	0.0131	1.9152	0.7975	0.0023	0.7342	−0.0106	−0.0674	−0.0075	0.008
0.15	0.6043	0.0155	1.9356	0.7971	0.0030	0.7345	−0.0104	−0.0673	−0.0071	0.0079
0.2	0.6050	0.0208	1.9557	0.7965	0.0078	0.7348	−0.0100	−0.0669	−0.0064	0.0076
0.25	0.6060	0.0367	1.9758	0.7957	0.0237	0.7352	−0.0093	−0.0656	−0.0056	0.0071
0.3	0.6083	0.0877	1.9957	0.7939	0.0754	0.7358	−0.0083	−0.0612	−0.0045	0.0064
0.35	0.6144	0.2526	2.0154	0.7892	0.2438	0.7364	−0.0063	−0.0470	−0.0032	0.0049
0.4	0.6328	0.7881	2.0351	0.7746	0.7913	0.7371	−0.0015	−0.0007	−0.0017	0.0014

Table 6.2. Optimal Solution for Closed Switch

t (s)	x_1	x_2	x_3	x_4	u_1	u_2	λ_1	λ_2	λ_3	λ_4
0.4	0.6328	0.7881	2.0351	0.7746	0.0847	0.6829	−0.2820	−0.0604	−0.1102	0.1466
0.405	0.6479	0.1119	2.0234	0.7622	0.7436	0.6659	−0.3217	−0.0048	−0.1443	0.1874
0.5	0.6474	−0.0941	2.0125	0.7626	0.9444	0.6666	−0.3003	0.0122	−0.1427	0.178
0.55	0.6427	−0.1433	2.0020	0.7666	0.9927	0.6728	−0.2632	0.0163	−0.1305	0.1547
0.6	0.6371	−0.142	1.9924	0.7712	0.9920	0.6799	−0.2247	0.0162	−0.1161	0.1298
0.605	0.6319	−0.126	1.9831	0.7755	0.9773	0.6868	−0.1892	0.015	−0.1023	0.1069
0.7	0.6274	−0.1071	1.9760	0.7792	0.9600	0.6932	−0.1575	0.0135	−0.0897	0.0868
0.705	0.6235	−0.0889	1.9689	0.7822	0.9436	0.6989	−0.1295	0.0121	−0.0783	0.0692
0.8	0.6204	−0.0726	1.9626	0.7847	0.9291	0.7040	−0.1048	0.0109	−0.0680	0.0538
0.805	0.6178	−0.0585	1.9568	0.7868	0.9162	0.7087	−0.0829	0.0098	−0.0586	0.0402
0.9	0.6157	−0.0475	1.9516	0.7884	0.9036	0.7129	−0.0635	0.0088	−0.0501	0.0281
0.905	0.6140	−0.0425	1.9469	0.7898	0.8879	0.7167	−0.0466	0.0074	−0.0426	0.0176
1.0	0.6122	−0.0527	1.9426	0.7912	0.8594	0.7196	−0.0332	0.0050	−0.0368	0.0094
1.05	0.6093	−0.1051	1.9386	0.7934	0.7899	0.7206	−0.0269	−0.0009	−0.0348	0.0063

the machines, i.e., no time constants are considered for either the governor or the voltage regulator.

The two machines are operating at steady state, and one of the machines is suddenly subjected to a torque pulse of 0.5 s duration. Thus, generally, the machines will experience a disturbance which could lead to loss of synchronism. It is desired to find optimally the controls that will minimize a cost functional, quadratic in the various state and control variables.

The controls are modeled as linear combinations of the state variables, but the machine equations are not linearized. Therefore, the model parameters, which are assumed to be time-independent, will generally depend on the strength and duration of the pulse, contrary to the linearized machine models where the equations themselves are linearized. Here, a different set of feedback parameters is needed to counteract different pulse strengths. This linear feedback model is considered, because it simplifies the formulation of the model as compared to the model used in Section 6.2.

The optimizing equations for the system are obtained by using the minimum-norm formulation of functional analysis. Then, using gradient

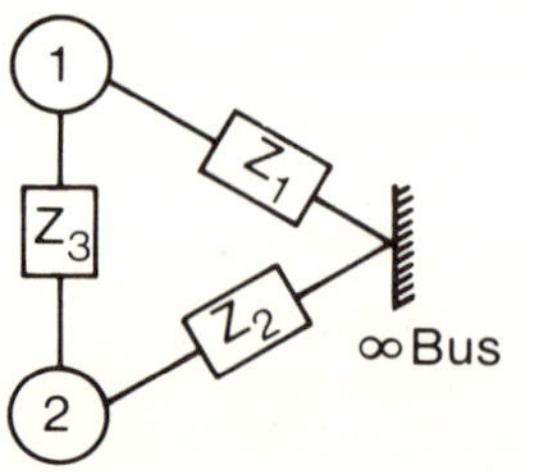

Figure 6.1. System considered.

descent methods (Ref. 6.10), the optimal feedback parameters are obtained by solving the nonlinear, two-point boundary-value problem on the digital computer.

6.3.1. Differential Equations of the System and Their Simplifications

The system considered is equivalent to that shown in Fig. 6.2; from Kirchhoff's current law and basic voltage–current relations, we have

$$I_1 = (V_1 - V_B)/z_1 + (V_1 - V_2)/z_3 \tag{6.60}$$

$$I_2 = (V_2 - V_1)/z_3 + (V_2 - V_B)/z_2 \tag{6.61}$$

or

$$\begin{bmatrix} I_1 \\ I_2 \end{bmatrix} = \begin{bmatrix} 1/z_1 + 1/z_3 & -1/z_3 \\ -1/z_3 & 1/z_2 + 1/z_3 \end{bmatrix} \begin{bmatrix} V_1 \\ V_2 \end{bmatrix} - \begin{bmatrix} V_{/}z_1 \\ V_B/z_2 \end{bmatrix} \tag{6.62}$$

where the reference axes D, Q refer to the infinite bus; all the voltages and currents are referred to the reference axis, i.e.,

$$I_1 = i_{D1} + ji_{Q1}, \qquad V_1 = v_{D1} + jv_{Q1} \tag{6.63}$$

$$I_2 = i_{D2} + ji_{Q2}, \qquad V_2 = v_{D2} + jv_{Q2} \tag{6.64}$$

which are related to the machines by the torque angles δ_1 and δ_2. We have, from the phasor diagram (Fig. 6.3),

$$I_1 = I_{m1}e^{j\delta_1}, \qquad V_1 = V_{m1}e^{j\delta_1} \tag{6.65}$$

$$I_2 = I_{m2}e^{j\delta_2}, \qquad V_2 = V_{m2}e^{j\delta_2} \tag{6.66}$$

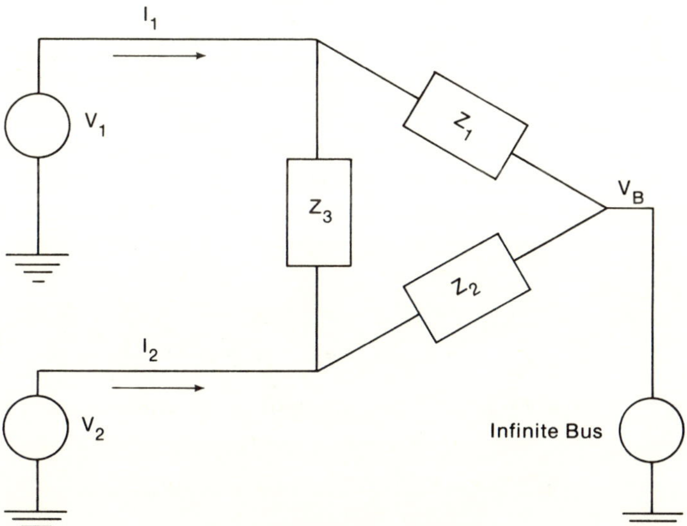

Figure 6.2. Equivalent circuit.

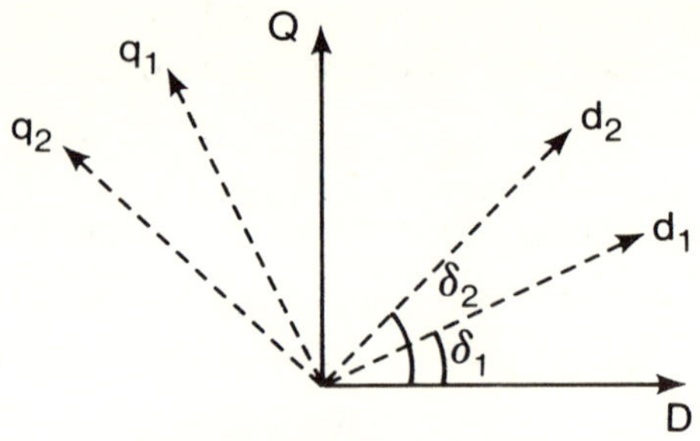

Figure 6.3. Phasor diagram.

where $I_{m1}, I_{m2}, V_{m1}, V_{m2}$ are phasors with respect to each machine axis (d_1, q_1) and (d_2, q_2).

Since the infinite bus serves as the reference junction, and if the bus voltage is to be in the form

$$v_B(t) = U_B \sin(\omega_0 t) \tag{6.67}$$

then V_B is on the imaginary axis, i.e.,

$$V_B = jU_B = j\sqrt{2}U_{\text{rms}} \tag{6.68}$$

Now, in the machine reference, we have

$$\begin{bmatrix} I_{m1}e^{j\delta_1} \\ I_{m2}e^{j\delta_2} \end{bmatrix} = \begin{bmatrix} \dfrac{1}{z_1}+\dfrac{1}{z_3} & -\dfrac{1}{z_3} \\ -\dfrac{1}{z_3} & \dfrac{1}{z_2}+\dfrac{1}{z_3} \end{bmatrix} \begin{bmatrix} V_{m1}e^{j\delta_1} \\ V_{m2}e^{j\delta_2} \end{bmatrix} - j \begin{bmatrix} (1/z_1)\sqrt{2}U_{\text{rms}} \\ (1/z_2)\sqrt{2}U_{\text{rms}} \end{bmatrix} \tag{6.69}$$

For reactances only, i.e.,

$$z_1 = jx_1 = j/y_1, \qquad z_2 = jx_2 = j/y_2, \qquad z_3 = jx_3 = j/y_3 \tag{6.70}$$

where y_1, y_2, y_3 are admittances, we have

$$\begin{bmatrix} (i_{d1}+ji_{q1})e^{j\delta_1} \\ (i_{d2}+ji_{q2})e^{j\delta_2} \end{bmatrix} = -j \begin{bmatrix} y_1+y_3 & -y_3 \\ -y_3 & y_2+y_3 \end{bmatrix} \begin{bmatrix} (u_{d1}+ju_{q1})e^{j\delta_1} \\ (u_{d2}+ju_{q2})e^{j\delta_2} \end{bmatrix} - \begin{bmatrix} y_1\sqrt{2}U_{\text{rms}} \\ y_2\sqrt{2}U_{\text{rms}} \end{bmatrix} \tag{6.71}$$

These are two complex equations with eight unknowns. We also have the current equations (Ref. 1)

$$i_{d1} = -\alpha_1 u_{q1} - \beta_1 x_1, \qquad i_{q1} = \gamma_1 u_{d1} \tag{6.72}$$

$$i_{d2} = -\alpha_2 u_{q2} - \beta_2 x_2, \qquad i_{q2} = \gamma_2 u_{d2} \tag{6.73}$$

so we have altogether eight real equations and eight unknowns. Substituting equations (6.72) and (6.73) into equations (6.71) we obtain two complex equations (i.e., four real equations) to determine the unknowns u_{d1}, u_{q1}, u_{d2}, u_{q2}:

$$\begin{bmatrix}\{-(\alpha_1 u_{q1}+\beta_1 x_1)+j\gamma_1 u_{d1}\}e^{j\delta_1}\\ \{-(\alpha_2 u_{q2}+\beta_2 x_2)+j\gamma_2 u_{d2}\}e^{j\delta_2}\end{bmatrix} = -j\begin{bmatrix} y_1+y_3 & -y_3\\ -y_3 & y_2+y_3\end{bmatrix}\begin{bmatrix}(u_{d1}+ju_{q1})e^{j\delta_1}\\ (u_{d2}+ju_{q2})e^{j\delta_2}\end{bmatrix} - \begin{bmatrix} y_1\sqrt{2}U_{\text{rms}}\\ y_2\sqrt{2}U_{\text{rms}}\end{bmatrix} \tag{6.74}$$

Once we solve for these, i.e., u_{d1}, u_{q1}, u_{d2}, u_{q2}, we can substitute into the torque and flux equations of the two-machine system, i.e.,

$$d^2\delta_1/dt^2 = (\omega_0/2H_1)\{\tfrac{1}{2}[(\alpha_1-\gamma_1)u_{d1}y_{q1}+\beta_1 u_{d1}x_1] - M_{t1} - K_{d1}(d\delta_1/dt)\} \tag{6.75}$$

$$dx_1/dt = \omega_0 u_{f1} - \omega_0 R_{f1}\{[x_{md1}/(x_{d1}x_{fd1}-x_{md1}^2)]u_{q1} + [x_{d1}/(x_{d1}x_{fd1}-x_{md1}^2)]x_1\} \tag{6.76}$$

and

$$d^2\delta_2/dt^2 = (\omega_0/2H_1)\{\tfrac{1}{2}[(\alpha_2-\gamma_2)u_{d2}u_{q2}+\beta_2 u_{d2}x_2] - M_{t2} - K_{d2}(d\delta_2/dt)\} \tag{6.77}$$

$$dx_2/dt = \omega_0 u_{f2} - \omega_0 R_{f2}\{[x_{md2}/(x_{d2}x_{fd2}-x_{md2}^2)]u_{q2} + [x_{d2}/(x_{d2}x_{fd2}-x_{md2}^2)]x_2\} \tag{6.78}$$

Hence, the differential equations of the two-machine system can be obtained.

Equation (6.74) can be rewritten in component form as follows:

$$[-(\alpha_1 u_{q1}+\beta_1 x_1)+j\gamma_1 u_{d1}+j(y_1+y_3)(u_{d1}+ju_{q1})]e^{j\delta_1} = jy_3(u_{d2}+ju_{q2})e^{j\delta_2} - y_1\sqrt{2}U_{\text{rms}} \tag{6.79}$$

$$[-(\alpha_2 u_{q2}+\beta_2 x_2)+j\gamma_2 u_{d2}+j(y_2+y_3)(u_{d2}+ju_{q2})]e^{j\delta_2} = jy_3(u_{d1}+ju_{q1})e^{j\delta_1} - y_2\sqrt{2}U_{\text{rms}} \tag{6.80}$$

or

$$[-(\alpha_1+y_1+y_3)u_{q1}-\beta_1 x_1+j(\gamma_1+y_1+y_3)u_{d1}]e^{j\delta_1} = (jy_3u_{d2}-y_3u_{q2})e^{j\delta_2} - y_1\sqrt{2}U_{\text{rms}} \tag{6.81}$$

$$[-(\alpha_2+y_2+y_3)u_{q2}-\beta_2 x_2+j(\gamma_2+y_2+y_3)]u_{d2}e^{j\delta_2} = (jy_3u_{d1}-y_3u_{q1})e^{j\delta_1} - y_2\sqrt{2}U_{\text{rms}} \tag{6.82}$$

Separating into real and imaginary components we have

$$-(\alpha_1 + y_1 + y_3)u_{q1} - \beta_1 x_1 = -y_3 u_{q2}\cos(\delta_2 - \delta_1) - y_3 u_{d2}\sin(\delta_2 - \delta_1) - y_1\sqrt{2}U_{\text{rms}}\cos\delta_1 \quad (6.83)$$

$$(\gamma_1 + y_1 + y_3)u_{d1} = y_3 u_{d2}\cos(\delta_2 - \delta_1) - y_3 u_{q2}\sin(\delta_2 - \delta_1) + y_1\sqrt{2}U_{\text{rms}}\sin\delta_1 \quad (6.84)$$

$$-(\alpha_2 + y_2 + y_3)u_{q2} - \beta_2 x_2 = -y_3 u_{q1}\cos(\delta_2 - \delta_1) + y_3 u_{d1}\sin(\delta_2 - \delta_1) - y_2\sqrt{2}U_{\text{rms}}\cos\delta_2 \quad (6.85)$$

$$(\gamma_2 + y_2 + y_3)u_{d2} = y_3 u_{d1}\cos(\delta_2 - \delta_1) + y_3 u_{q1}\sin(\delta_2 - \delta_1) + y_2\sqrt{2}U_{\text{rms}}\sin\delta_2 \quad (6.86)$$

Substituting equations (6.83) and (6.84) into equations (6.85) and (6.86) we have

$$\begin{aligned}
-(\alpha_2 + y_2 + y_3)u_{q2} - \beta_2 x_2
&= [y_3\cos(\delta_2 - \delta_1)/(\alpha_1 + y_1 + y_3)][\beta_1 x_1 - y_3 u_{q2}\cos(\delta_2 - \delta_1) \\
&\quad - y_3 u_{d2}\sin(\delta_2 - \delta_1) - y_1\sqrt{2}U_{\text{rms}}\cos\delta_1] - y_2\sqrt{2}U_{\text{rms}}\cos\delta_2 \\
&\quad + [y_3\sin(\delta_2 - \delta_1)/(\gamma_1 + y_1 + y_3)] \\
&\quad \times [y_3 u_{d2}\cos(\delta_2 - \delta_1) - y_3 u_{q2}\sin(\delta_2 - \delta_1) + y_1\sqrt{2}U_{\text{rms}}\sin\delta_1]
\end{aligned} \quad (6.87)$$

$$\begin{aligned}
(\gamma_2 + y_2 + y_3)u_{d2}
&= [(y_3\cos(\delta_2 - \delta_1)/(\gamma_1 + y_1 + y_3)][y_3 u_{d2}\cos(\delta_2 - \delta_1) \\
&\quad - y_3 u_{q2}\sin(\delta_2 - \delta_1) + y_1\sqrt{2}U_{\text{rms}}\sin\delta_1] \\
&\quad - [y_3\sin(\delta_2 - \delta_1)/(\alpha_1 + y_1 + y_3)] \\
&\quad \times [\beta_1 x_1 - y_3 u_{q2}\cos(\delta_2 - \delta_1) - y_3 u_{d2}\sin(\delta_2 - \delta_1) - y_1\sqrt{2}U_{\text{rms}}\cos\delta_1] \\
&\quad + y_2\sqrt{2}U_{\text{rms}}\sin\delta_2
\end{aligned} \quad (6.88)$$

Solving the linear equations (6.87) and (6.88) for u_{d2} and u_{q2}, we have

$$\begin{aligned}
u_{d2} &= \{1/[A + B\cos^2(\delta_2 - \delta_1)]\} \\
&\quad \times [E_1F_1\sin\delta_2 + (E_1F_2 + E_2F_1 + E_2F_2)\sin\delta_1\cos(\delta_2 - \delta_1) \\
&\quad - E_1F_3F_4\cos\delta_2\sin(\delta_2 - \delta_1) + \beta_1 E_1F_3x_1\sin(\delta_2 - \delta_1) \\
&\quad + \beta_2E_2x_2\sin(\delta_2 - \delta_1)\cos(\delta_2 - \delta_1)]
\end{aligned} \quad (6.89)$$

$$\begin{aligned}
u_{q2} &= \{1/[A + B\cos^2(\delta_2 - \delta_1)]\} \\
&\quad \times [-E_3F_1\cos\delta_2 + (E_2F_3F_4 + E_3F_3F_4 - E_2F_1)\cos\delta_1\cos(\delta_2 - \delta_1)
\end{aligned}$$

$$+ E_3\beta_2 x_2 + E_3 F_2 \sin\delta_1 \sin(\delta_2 - \delta_1) + E_2\beta_2 x_2 \cos^2(\delta_2 - \delta_1)$$

$$- (E_2 + E_3) F_3 \beta_1 x_1 \cos(\delta_2 - \delta_1)] \tag{6.90}$$

where

$$A = [-(\alpha_2 + y_2 + y_3) + y_3^2/(\gamma_1 + y_1 + y_3)]$$

$$\times [(\gamma_2 + y_2 + y_3) - y_3^2/(\alpha_1 + y_1 + y_3)] \tag{6.91}$$

$$B = (\gamma_2 - \alpha_2)[1/(\alpha_1 + y_1 + y_3) - 1/(\gamma_1 + y_1 + y_3)]y_3^2 \tag{6.92}$$

$$E_1 = -(\alpha_2 + y_2 + y_3) + y_3^2/(\gamma_1 + y_1 + y_3) \tag{6.93}$$

$$E_2 = [1/(\alpha_1 + y_1 + y_3) - 1/(\gamma_1 + y_1 + y_3)]y_3^2 \tag{6.94}$$

$$E_3 = (\alpha_2 + y_2 + y_3) - y_3^2/(\alpha_1 + y_1 + y_3) \tag{6.95}$$

$$F_1 = \sqrt{2} y_2 U_{\text{rms}} \tag{6.96}$$

$$F_2 = \sqrt{2} y_1 y_3 U_{\text{rms}}/(\gamma_1 + y_1 + y_3) \tag{6.97}$$

$$F_3 = -y_3/(\alpha_1 + y_1 + y_3) \tag{6.98}$$

$$F_4 = \sqrt{2} y_1 U_{\text{rms}} \tag{6.99}$$

u_{d2} and u_{q2} are then substituted in equations (6.83) and (6.84). Solving for u_{d1} and u_{q1}, we have

$$u_{d1} = \{1/[A^* + B^* \cos^2(\delta_2 - \delta_1)]\}$$

$$\times [E_1^* F_1^* \sin\delta_1 - (E_1^* F_2^* + E_2^* F_1^* + E_2^* F_2^*) \sin\delta_2 \cos(\delta_1 - \delta_2)$$

$$- E_1^* F_3^* F_4^* \cos\delta_2 \sin(\delta_1 - \delta_2) - \beta_2 E_1^* F_3^* x_2 \sin(\delta_1 - \delta_2)$$

$$+ \beta_1 E_2^* x_1 \sin(\delta_1 - \delta_2) \cos(\delta_1 - \delta_2)] \tag{6.100}$$

$$u_{q1} = \{1[A^* + B^* \cos^2(\delta_2 - \delta_1)]\}[-E_3^* F_1^* \cos\delta_1$$

$$+ (E_2^* F_3^* F_4^* + E_3^* F_3^* F_4^* - E_2^* F_1^*) \cos\delta_2 \cos(\delta_1 - \delta_2)$$

$$+ E_3^* \beta_1 x_1 + E_3^* F_2^* \sin\delta_2 \sin(\delta_1 - \delta_2)$$

$$+ E_2^* \beta_1 x_1 \cos^2(\delta_1 - \delta_2) - (E_2^* + E_3^*) F_3^* \beta_2 x_2 \cos(\delta_1 - \delta_2)] \tag{6.101}$$

where

$$A^* = [-(\alpha_1 + y_1 + y_3) + y_3^2/(\gamma_2 + y_2 + y_3)]$$

$$\times [(\gamma_1 + y_1 + y_3) - y_3^2/(\alpha_2 + y_2 + y_3)] \tag{6.102}$$

$$B^* = (\gamma_1 - \alpha_1)[(1/(\alpha_2 + y_2 + y_3) - 1/(\gamma_2 + y_2 + y_3)]y_3^2 \tag{6.103}$$

$$E_1^* = -(\alpha_1 + y_1 + y_3) + y_3^2/(\gamma_2 + y_2 + y_3) \tag{6.104}$$

$$E_2^* = [1/(\alpha_2 + y_2 + y_2) - 1/(\gamma_2 + y_2 + y_3)]y_3^2 \tag{6.105}$$

$$E_3^* = (\gamma_1 + y_1 + y_3) - y_3^2/(\alpha_2 + y_2 + y_3) \tag{6.106}$$

$$F_1^* = \sqrt{2}y_1 U_{\text{rms}} \tag{6.107}$$

$$F_2^* = \sqrt{2}y_2 y_3 U_{\text{rms}}/(\gamma_2 + y_2 + y_3) \tag{6.108}$$

$$F_3^* = -y_3/(\alpha_2 + y_2 + y_3) \tag{6.109}$$

$$F_4^* = \sqrt{2}y_2 U_{\text{rms}} \tag{6.110}$$

By convention, the changes

$$\begin{aligned} (\delta_1, \delta_2) &\rightarrow ((\delta_1' = -\delta_1), (\delta_2' = -\delta_2)) \\ (x_1, x_2) &\rightarrow ((x_1' = -x_1, x_2' = -x_2)) \end{aligned} \tag{6.111}$$

are made to get generator action for $\delta_i' > 0$. So, the following differential equations for the two-machine system are obtained:

$$d\delta_1'/dt = \eta_1' \tag{6.112}$$

$$d\eta_1'/dt = (\omega_0/2H_1)\{(1/2)(\gamma_1 - \alpha_1)u_{d1}u_{q1} + \beta_1 u_{d1}x_1'] + M_{t1} - K_{d1}\eta_1'\} \tag{6.113}$$

$$\begin{aligned} dx_1'/dt = -\omega_0 u_{f1} + \omega_0 R_{f1}\{[x_{md1}/(x_{d1}x_{fd1} - x_{md1}^2)]u_{q1} \\ - [x_{d1}/(x_{d1}x_{fd1} - x_{md1}^2)]x_1'\} \end{aligned} \tag{6.114}$$

$$d\delta_2'/dt = \eta_2' \tag{6.115}$$

$$d\eta_2'/dt = (\omega_0/2H_2)\{(1/2)[(\gamma_2 - \alpha_2)u_{d2}u_{q2} + \beta_2 u_{d2}x_2'] + M_{t2} - K_{d2}\eta_2'\} \tag{6.116}$$

$$\begin{aligned} dx_2'/dt = -\omega_0 u_{f2} + \omega_0 R_{f2}\{[x_{md2}/(x_{d2}x_{fd2} - x_{md2}^2)]u_{q2} \\ - [x_{d2}/(x_{d2}x_{fd2} - x_{md2}^2)]x_2'\} \end{aligned} \tag{6.117}$$

and

$$\begin{aligned} u_{d1} = \{1/[A^* + B^* \cos^2(\delta_2' - \delta_1')]\} \\ \times [-E_1^*F_1^* \sin\delta_1' - (E_1^*F_2^* + E_2^*F_1^* + E_2^*F_2^*)\sin\delta_2' \cos(\delta_1' - \delta_2') \\ + E_1^*F_3^*F_4^* \cos\delta_2' \sin(\delta_1' - \delta_2') + \beta_2 E_1^*F_3^*x_2' \sin(\delta_1' - \delta_2') \\ + \beta_1 E_2^* x_1' \sin(\delta_1' - \delta_2')\cos(\delta_1' - \delta_2')] \end{aligned} \tag{6.118}$$

$$\begin{aligned} u_{q1} = \{1/[A^* + B^* \cos^2(\delta_2' - \delta_1)]\} \\ \times [-E_3^*F_1^* \cos\delta_1' + (E_2^*F_3^*F_4^* + E_3^*F_3^*F_4^* - E_2^*F_1^*)\cos\delta_2' \cos(\delta_2' - \delta_2') \\ - E_3^*\beta_1 x_1' + E_3^*F_2^* \sin\delta_2' \sin(\delta_1' - \delta_2') - E_2^*\beta_1 x_2' \cos^2(\delta_1' - \delta_2') \\ + (E_2^* + E_3^*)F_3^*\beta_2 x_2' \cos(\delta_1' - \delta_2')] \end{aligned} \tag{6.119}$$

$$\begin{aligned} u_{d2} = \{1/[A + B\cos^2(\delta_2' - \delta_1')]\} \\ \times [-E_1F_1 \sin\delta_2' - (E_1F_2 + E_2F_1 + E_2F_2)\sin\delta_1' \cos(\delta_2' - \delta_1') \end{aligned}$$

$$+ E_1F_3F_4 \cos\delta_1' \sin(\delta_2'-\delta_1') + \beta_1E_1F_3x_1' \sin(\delta_2'-\delta_1')$$

$$+ \beta_2E_2x_2' \sin(\delta_2'-\delta_1') \cos(\delta_2'-\delta_1')] \tag{6.120}$$

$$\begin{aligned} u_{q1} = &\{1/[A+B\cos^2(\delta_2'-\delta_1')]\} \\ &\times [E_3F_1\cos\delta_2' + (E_2F_3F_4+E_3F_3F_4-E_2F_1)\cos\delta_1'\cos(\delta_2'-\delta_1') \\ &- E_3\beta_2x_2' + E_3F_2\sin\delta_1'\sin(\delta_2'-\delta_1') - E_2\beta_2x_2'\cos^2(\delta_2'-\delta_1') \\ &+ (E_2+E_3)F_3\beta_1x_1'\cos(\delta_2'-\delta_1')] \end{aligned} \tag{6.121}$$

If $y_3 = 0$, a system of two separated machines connected to the infinite bus is obtained and equation (6.74) becomes

$$\begin{bmatrix} (-(\alpha_1u_{q1}+\beta_1x_1)+j\gamma_1u_{d1})e^{j\delta_1} \\ (-(\alpha_2u_{q2}+\beta_2x_2)+j\gamma_2u_{d2})e^{j\delta_2} \end{bmatrix} = -j\begin{bmatrix} y_1 & 0 \\ 0 & y_2 \end{bmatrix}\begin{bmatrix} (u_{d1}+ju_{q1})e^{j\delta_1} \\ (u_{d2}+ju_{q2})e^{j\delta_2} \end{bmatrix} - \begin{bmatrix} y_1\sqrt{2}U_{\rm rms} \\ y_2\sqrt{2}U_{\rm rms} \end{bmatrix} \tag{6.122}$$

or

$$\{-[(y_1+\alpha_1)u_{q1}+\beta_1x_1]+j(\gamma_1+y_1)u_{d1}\}e^{j\delta_1} = -y_1\sqrt{2}U_{\rm rms} \tag{6.123}$$

$$\{-[(y_2+\alpha_2)u_{q2}+\beta_2x_2]+j(\gamma_2+y_2)u_{d2}\}e^{j\delta_2} = -y_2\sqrt{2}U_{\rm rms} \tag{6.124}$$

Hence, the two machines separate, and we can look at equation (6.123) or equation (6.124) separately. Consider equation (6.123), and set

$$z_1 = jx_e \tag{6.125}$$

Separating real and imaginary parts in equation (6.123), we have

$$(y_1+\alpha_1)u_{q1}+\beta_1x_1 = y_1\sqrt{2}U_{\rm rms}\cos\delta_1 \tag{6.126}$$

$$(\gamma_1+y_1)u_{d1} = y_1\sqrt{2}U_{\rm rms}\sin\delta_1 \tag{6.127}$$

and hence

$$\begin{aligned} u_{q1} &= -[\beta_1/(y_1+\alpha_1)]x_1 + [y_1\sqrt{2}U_{\rm rms}/(y_1+\alpha_1)]\cos\delta_1 \\ &= [\beta_1/(y_1+\alpha_1)]x_1' + [y_1\sqrt{2}U_{\rm rms}/(y_1+\alpha_1)]\cos\delta_1' \end{aligned} \tag{6.128}$$

$$u_{d1} = [y_1\sqrt{2}U_{\rm rms}(\gamma_1+y_1)]\sin\delta_1 = -[y_1\sqrt{2}U_{\rm rms}/(\gamma_1+y_1)]\sin\delta_1' \tag{6.129}$$

Substitute equations (6.128) and (6.129) into equations (6.112)-(6.117), and note that

$$\alpha_1 = x_{fd1}/(x_{d1}x_{fd1}-x_{md1}^2) \tag{6.130}$$

$$\beta_1 = x_{md1}/(x_{d1}x_{fd1}-x_{md1}^2) \tag{6.131}$$

$$\gamma_1 = 1/x_{q1} \tag{6.132}$$

$$y_1 = 1/x_{e1} \tag{6.133}$$

$$x'_{q1} = x_{q1} + x_{e1} \tag{6.134}$$

$$x'_{d1} = x_{d1} + x_{e1} \tag{6.135}$$

We have

$$d\delta_1/dt = \eta'_1 \tag{6.136}$$

$$d\eta'_1/dt = (1/M_1)(M_t - S_4 \sin\delta'_1 \cos\delta'_1 - S_5 x'_1 \sin\delta'_1 - K_{d1}\eta'_1 \tag{6.137}$$

$$dx'_1/dt = -\omega_0 u_{f1} - Ax'_1 + C_1 \cos\delta'_1 \tag{6.138}$$

where

$$M_1 = 2H_1/\omega_0 \tag{6.139}$$

$$S_4 = [-U^2_{\text{rms}}(x'_{q1} - x'_{d1} + m_{md1}/x_{fd1})]/[x'_{q1}(x'_{d1} - x^2_{md1}/x_{fd1})] \tag{6.140}$$

$$S_5 = (1/2)U_{\text{rms}}(x_{md1}/x_{fd1})[1/(x'_{d1} - x^2_{md1}/x_{fd1})] \tag{6.141}$$

$$A = (x'_{d1}/x_{fd1})[1/(x'_{d1} - x^2_{md1}/x_{fd1})]\omega_0 R_{f1} \tag{6.142}$$

$$C = [(2U_{\text{rms}})^{1/2}\omega_0 R_{f1} x_{md1}]/[x_{fd1}(x'_{d1} - x_{md1}/x_{fd1})] \tag{6.143}$$

Note that equations (6.136)–(6.138) are identical to the dynamical equations for a one-machine system given in (6.1)–(6.3).

If we set

$$y_3 = 0 \quad \text{and} \quad y_1 = 1/x_{e1}$$

in equations (6.102)–(6.110), using equations (6.112)–(6.121), the same dynamical equations for the one-machine system, equations (6.136)–(6.138) are also obtained.

Equations (6.102)–(6.110) are somewhat awkward, because they make the system nonpolynomic in view of the terms

$$1/[A + B\cos^2(\delta'_2 - \delta'_1)] \quad \text{and} \quad 1/[A^* + B^*\cos^2(\delta'_2 - \delta'_1)]$$

Note that

$$A = A^* \simeq -140, \qquad B = B^* \simeq 3$$

in practice. Therefore,

$$1/[A^* + B^*\cos^2(\delta'_2 - \delta'_1)] \simeq 1/(A^* + B^*) \tag{6.144}$$

$$1/[A + B\cos^2(\delta'_2 - \delta'_1)] \simeq 1/(A + B) \tag{6.145}$$

Furthermore, only the deviations of the state variables about steady-state values are considered. Therefore, we assume that, for the system at hand,

$$\sin(\delta'_2 - \delta'_1) \simeq 0, \qquad \cos(\delta'_2 - \delta'_1) \simeq 1 \tag{6.146}$$

Using

$$\sin\delta_1'\cos(\delta_2'-\delta_1')+\cos\delta_1'\sin(\delta_2'-\delta_1')=\sin\delta_2' \tag{6.147}$$

$$\cos\delta_1'\cos(\delta_1'-\delta_2')+\sin\delta_1'\sin(\delta_1'-\delta_2')=\cos\delta_2' \tag{6.148}$$

we obtain, from equations (6.118)-(6.121),

$$u_{d1}\simeq R_1^*\sin\delta_1'+R_2^*\sin\delta_2' \tag{6.149}$$

$$u_{q1}\simeq S_1^*\cos\delta_1'+S_2^*\cos\delta_2'+S_3^*x_1'+S_4^*x_2' \tag{6.150}$$

$$u_{d2}\simeq R_1\sin\delta_1'+R_2\sin\delta_2' \tag{6.151}$$

$$u_{q2}\simeq S_1\cos\delta_1'+S_2\cos\delta_2'+S_3x_1'+S_4x_2' \tag{6.152}$$

where

$$R_1^*=E_1^*(F_3^*F_4^*-F_1^*)/(A^*+B^*) \tag{6.153}$$

$$R_2^*=-(E_1^*F_2^*+E_2^*F_1^*+E_2^*F_2^*+E_1^*F_3^*F_4^*)/(A^*+B^*) \tag{6.154}$$

$$S_1^*=-E_3^*(F_1^*+F_2^*)/(A^*+B^*) \tag{6.155}$$

$$S_2^*=-(E_2^*F_3^*F_4^*+E_3^*F_3^*F_4^*+E_3^*F_2^*-E_2^*F_1^*)/(A^*+B^*) \tag{6.156}$$

$$S_3^*=-\beta_1(E_2^*+E_3^*)/(A^*+B^*) \tag{6.157}$$

$$S_4^*=\beta_2F_3^*(E_2^*+E_3^*)/(A^*+B^*) \tag{6.158}$$

$$R_1=-(E_1F_2+E_2F_1+E_2F_2+E_1F_3F_4)/(A+B) \tag{6.159}$$

$$R_2=E_1(F_3F_4-F_1)/(A+B) \tag{6.160}$$

$$S_1=(E_2F_3F_4+E_3F_3F_4+E_3F_2-E_2F_1)/(A+B) \tag{6.161}$$

$$S_2=-E_3(F_1+F_2)/(A+B) \tag{6.162}$$

$$S_3=\beta_1F_3(E_2+E_3)/(A+B) \tag{6.163}$$

$$S_4=-\beta_2(E_2+E_3)/(A+B) \tag{6.164}$$

6.3.2. Mathematical Model

As indicated in Section 6.2, M_t and $-\omega_0u_f$ are considered to be controls, and the following transformations are introduced:

$$x_1=\sin\delta_1' \tag{6.165}$$

$$x_2=\eta_1' \tag{6.166}$$

$$x_3=x_1' \tag{6.167}$$

$$x_4=\cos\delta_1' \tag{6.168}$$

$$x_5=\sin\delta_2' \tag{6.169}$$

$$x_6=\eta_2' \tag{6.170}$$

$$x_7 = x_2' \tag{6.171}$$

$$x_8 = \cos \delta_2' \tag{6.172}$$

Then, equations (6.112)-(6.117) become

$$dx_1/dt = x_2 x_4 \tag{6.173}$$

$$dx_2/dt = a_1 u_{d1} u_{q1} + b_1 u_{d1} x_3 + (1/M_1) U_1 - K_{d1}^* x_2 \tag{6.174}$$

$$dx_3/dt = V_1 + C_1 u_{q1} - d_1 x_3 \tag{6.175}$$

$$dx_4/dt = -x_1 x_2 \tag{6.176}$$

$$dx_5/dt = x_6 x_8 \tag{6.177}$$

$$dx_6/dt = a_2 u_{d2} u_{q2} + b_2 u_{d2} x_7 + (1/M_2) U_2 - K_{d2}^* x_6 \tag{6.178}$$

$$dx_7/dt = V_2 + C_2 u_{q2} - d_2 x_7 \tag{6.179}$$

$$dx_8/dt = -x_5 x_6 \tag{6.180}$$

where

$$U_1 = M_{t1}, \qquad V_1 = -\omega_0 u_{f1} \tag{6.181}$$

$$U_2 = M_{t2}, \qquad V_2 = -\omega_0 u_{f2} \tag{6.182}$$

$$u_{d1} = R_1^* x_1 + R_2^* x_5, \qquad u_{q1} = S_1^* x_4 + S_2^* x_8 + S_3^* x_3 + S_4^* x_7 \tag{6.183}$$

$$u_{d2} = R_1 x_1 + R_2 x_5, \qquad u_{q2} = S_1 x_4 + S_2 x_8 + S_3 x_3 + S_4 x_7 \tag{6.184}$$

$$M_1 = 2H_1/\omega_0, \qquad M_2 = 2H_2/\omega_0 \tag{6.185}$$

$$a_1 = (1/2M_1)(\gamma_1 - \alpha_1), \qquad b_1 = (1/2M_1)\beta_1 \tag{6.186}$$

$$c_1 = \omega_0 R_{f1} x_{md1}/(x_{d1} x_{fd1} - x_{md1}^2), \quad d_1 = \omega_0 R_{f1} x_{d1}/(x_{d1} x_{fd1} - x_{md1}^2) \tag{6.187}$$

$$K_{d1}^* = K_{d1}/M_1, \qquad K_{d2}^* = K_{d2}/M_2 \tag{6.188}$$

$$a_2(1/2M_2)(\gamma_2 - \alpha_2), \qquad b_2 = (1/2M_2)\beta_2 \tag{6.189}$$

$$c_2 = \omega_0 R_{f2} x_{md2}/(x_{d2} x_{fd2} - x_{md2}^2), \quad d_2 = \omega_0 R_{f2} x_{d2}/(x_{d2} x_{fd2} - x_{md2}^2) \tag{6.190}$$

Deviations from the steady-state values are

$$y_1 = x_1 - x_1^s, \qquad y_2 = x_2 - x_2^s \tag{6.191}$$

$$y_3 = x_3 - x_3^s, \qquad y_4 = x_4 - x_4^s \tag{6.192}$$

$$y_5 = x_5 - x_5^s, \qquad y_6 = x_6 - x_6^s \tag{6.193}$$

$$y_7 = x_7 - x_7^s, \qquad y_8 = x_8 - x_8^s \tag{6.194}$$

$$u_1 = U_1 - U_1^s, \qquad v_1 = V_1 - V_1^s \tag{6.195}$$

$$u_2 = U_2 - U_2^s, \qquad u_2 = V_2 - V_2^s \tag{6.196}$$

The relations among the steady-state values are as follows:

$$x_2^s = 0 \tag{6.197}$$

$$x_6^s = 0 \tag{6.198}$$

$$a_1 u_{d1}^s u_{q1}^s + b_1 x_3^s u_{d1}^s + (1/M_1) U_1^s = 0 \tag{6.199}$$

$$V_1^s + C_1 u_{q1}^s - d_1 x_3^s = 0 \tag{6.200}$$

$$a_2 u_{d2}^s u_{q2}^s + b_2 x_7^s u_{d2}^s + (1/M_2) U_2^s = 0 \tag{6.201}$$

$$V_2^s + C_2 u_{q2}^s - d_2 x_7^s = 0 \tag{6.202}$$

where

$$u_{d1}^s = R_1^* x_1^s + R_2^* x_5^s \tag{6.203}$$

$$u_{q1}^s = S_1^* x_4^s + S_2^* x_8^s + S_3^* x_3^s + S_4^* x_7^s \tag{6.204}$$

$$u_{d2}^s = R_1 x_1^s + R_2 x_5^s \tag{6.205}$$

$$u_{q2}^s = S_1 x_4^s + S_2 x_8^s + S_3 x_3^s + S_4 x_7^s \tag{6.206}$$

It is assumed that the disturbance is applied to one of the two machines and is modeled by introducing a torque pulse in the right-hand side of equations (6.173)-(6.180). The magnitude of the torque pulse K was increased in steps of 5% of its steady-state value for a duration of 0.5 s. Substituting equations (6.191)-(6.196) into equations (6.83) and using equations (6.197)-(6.202) we obtain the dynamical equations of deviation

$$dy_1/dt = x_4^s y_2 + y_2 y_4 \tag{6.207}$$

$$\begin{aligned} dy_2/dt = {} & a_1(u_{d1}^s u_{q1}^y + u_{q1}^s u_{d1}^y + u_{d1}^y u_{q1}^y) + b_1(x_3^s u_{d1}^y + u_{d1}^s y_3 + u_{d1}^y y_3) \\ & + (1/M_1)u_1 - K_{d1}^* y_2 + (1/M)K[u(t - t_0) - u(t - \tau)] \end{aligned} \tag{6.208}$$

$$dy_3/dt = v_1 + c_1 u_{q1}^y - d_1 y_3 \tag{6.209}$$

$$dy_4/dt = -x_1^s y_2 - y_1 y_2 \tag{6.210}$$

$$dy_5/dt = x_8^s y_6 + y_6 y_8 \tag{6.211}$$

$$\begin{aligned} dy_6/dt = {} & a_2(u_{d2}^s u_{q2}^y + u_{q2}^s u_{d2}^y + u_{d2}^y u_{q2}^y) \\ & + b_2(x_7^s u_{d2}^y + u_{d2}^s y_7 + u_{d2}^y y_7) + (1/M_2)u_2 - K_{d2}^* y_6 \end{aligned} \tag{6.212}$$

$$dy_7/dt = v_2 + c_2 u_{q2}^y - d_2 y_7 \tag{6.213}$$

$$dy_8/dt = -x_5^s y_6 - y_5 y_6 \tag{6.214}$$

where

$$u_{d1}^y = R_1^* y_1 + R_2^* y_5 \tag{6.215}$$

$$u_{q1}^y = S_1^* y_4 + S_2^* y_8 + S_3^* y_3 + S_4^* y_7 \tag{6.216}$$

$$u_{d2}^{y} = R_1 y_1 + R_2 y_5 \tag{6.217}$$

$$u_{q2}^{y} = S_1 y_4 + S_2 y_8 + S_3 y_3 + S_4 y_7 \tag{6.218}$$

$$u(t) = \begin{cases} 1, & t > 0 \\ 0, & t < 0 \end{cases} \tag{6.219}$$

For this problem, the control is determined to be of the form (Ref. 6.12)

$$u_1 = \sum_{i=1}^{8} e_i y_i, \qquad v_1 = \sum_{i=1}^{8} f_i y_i \tag{6.220}$$

$$u_2 = \sum_{i=1}^{8} g_i y_i, \qquad v_2 = \sum_{i=1}^{8} h_i y_i \tag{6.221}$$

where all the e_i's, f_i's, g_i's, h_i's are constants. However, the magnitude of these constraints will in general depend on the strength of the disturbance. This can be stated mathematically as

$$de_i/dt = df_i/dt = dg_i/dt = dh_i/dt = 0, \qquad i = 1, 2, \ldots, 8 \tag{6.222}$$

The control which takes the form (6.220) and (6.221) is called global control (Ref. 6.11), whereas for the special case (6.23) and (6.224), which is called local control, the form is

$$u_1 = \sum_{i=1}^{4} e_i y_i, \qquad v_1 = \sum_{i=1}^{4} f_i y_i \tag{6.223}$$

$$u_2 = \sum_{i=1}^{4} g_i y_{i+4}, \qquad v_2 = \sum_{i-1}^{4} h_i y_{i+4} \tag{6.224}$$

The cost functional to be minimized is

$$J = \int_{t_0}^{t_f} \left[\sum_{i=1}^{8} (k_i y_i^2 + m_i e_i^2 + n_i f_i^2 + p_i g_i^2 + q_i h_i^2) + l_1 u_1^2 + l_2 v_1^2 + l_3 u_2^2 + l_4 v_2^2 \right] dt \tag{6.225}$$

for the case of global control. For the case of local control, the cost functional has the form

$$J = \int_{t_0}^{t_f} \left[\sum_{i=1}^{8} k_i y_i^2 + \sum_{i=1}^{4} (m_i e_i^2 + n_i f_i^2 + p_i g_i^2 + q_i h_i^2) + l_1 u_1^2 + l_2 v_1^2 + l_3 u_2^2 + l_4 v_2^2 \right] dt \tag{6.226}$$

where $y_i(t_0)$, $i = 1, 2, \ldots, 8$, and the k_i's, m_i's, n_i's, p_i's, q_i's, l_i's are constants.

6.3.3. Mathematical Solution

The augmented cost functional is obtained by adjoining equations (6.207)-(6.214), (6.220), (6.221), and (6.222) to the cost functional via Lagrange multipliers. One obtains

$$J_1 = \int_{t_0}^{t_f} F_G \, dt \tag{6.227}$$

where

$$\begin{aligned}
F_G = & \sum_{i=1}^{8} (k_i y_i^2 + m_i e_i^2 + n_i f_i^2 + p_i g_i^2 + q_i h_i^2) + l_1 u_1^2 + l_2 v_1^2 + l_3 u_2^2 + l_4 v_2^2 \\
& + \mu_1 \left(u_1 - \sum_{i=1}^{8} e_i y_i \right) + \mu_2 \left(v_1 - \sum_{i=1}^{8} f_i y_i \right) \\
& + \mu_3 \left(u_2 - \sum_{i=1}^{8} g_i y_i \right) + \mu_4 \left(v_2 - \sum_{i=1}^{8} h_i y_i \right) \\
& + \sum_{i=1}^{8} \nu_i (de_i/dt) + \sum_{i=1}^{8} \nu_{i+8} (df_i/dt) \\
& + \sum_{i=1}^{8} \nu_{i+16} (dg_i/dt) + \sum_{i=1}^{8} \nu_{i+24} (dh_i/dt) + \lambda_1 (dy/dt - x_4^s y_2 - y_2 y_4) \\
& + \lambda_2 \{ dy_2/dt - a_1 (u_{d1}^s u_{l1}^y + u_{q1}^s u_{d1}^y + u_{d1}^y u_{q1}^y) \\
& - b_1 (x_3^s u_{d1}^y + u_{d1}^s y_3 + u_{d1}^y y_3) - (1/M_1) u_1 + K_{d1}^* y_2 \\
& - (1/M_1) K [u(t - t_0) - u(t - \tau)] \} \\
& + \lambda_3 (dy_3/dt - v_1 - c_1 u_{q1}^y + d_1 y_3) + \lambda_4 (dy_4/dt + x_1^s y_2 + y_1 y_2) \\
& + \lambda_5 (dy_5/dt - x_8^s y_6 - y_6 y_8) \\
& + \lambda_6 [dy_6/dt - a_2 (u_{d2}^s u_{q2}^y + u_{q2}^s u_{d2}^y + u_{d2}^y u_{q2}^y) \\
& - b_2 (x_7^s u_{d2}^y + u_{d2}^s y_7 + u_{d2}^y y_7) - (1/M_2) u_2 + K_{d2}^* y_6] \\
& + \lambda_7 (dy_7/dt - v_2 - c_2 u_{q2}^y + d_2 y_7) \\
& + \lambda_8 (dy_8/dt + x_5^s y_6 + y_5 y_6)
\end{aligned} \tag{6.228}$$

where $\mu_1, \mu_2, \mu_3, \mu_4, \nu_1, \nu_2, \ldots, \nu_{32}, \lambda_1, \lambda_2, \ldots, \lambda_8$ are Lagrange multipliers. Using the same procedures as in the previous work (Refs. 6.7 and 6.8), the augmented cost functional reduces to

$$J_1 = \int_{t_0}^{t_f} [X^T R X + L^T X] \, dt + \text{boundary terms} \tag{6.229}$$

where T denotes transpose,

$$X^T = (y_1 \sim y_8, e_1 \sim e_8, f_1 \sim f_8, g_1 \sim g_8, h_1 \sim h_8, u_1, v_1, u_2, v_2)$$

is a vector of order 44,

$$L^T = (L_1 \sim L_8, L_9 \sim L_{16}, L_{17} \sim L_{24}, L_{25} \sim L_{32}, L_{33} \sim L_{40}, L_{41} \sim L_{44})$$

is a vector of order 44, and R is a matrix of order 44×44 which has the partitioned form shown below:

$$R = \begin{bmatrix} R_{11} & -\frac{1}{2}\mu_1 E_8 & \frac{1}{2}\mu_2 E_8 & -\frac{1}{2}\mu_3 E_8 & -\frac{1}{2}\mu_4 E_8 & 0_{8\times 4} \\ -\frac{1}{2}\mu_1 E_8 & M_1 & 0_{8\times 8} & 0_{8\times 8} & 0_{8\times 8} & 0_{8\times 4} \\ -\frac{1}{2}\mu_2 E_8 & 0_{8\times 8} & N_1 & 0_{8\times 8} & 0_{8\times 8} & 0_{8\times 4} \\ -\frac{1}{2}\mu_3 E_8 & 0_{8\times 8} & 0_{8\times 8} & P_1 & 0_{8\times 8} & 0_{8\times 4} \\ -\frac{1}{2}\mu_4 E_8 & 0_{8\times 8} & 0_{8\times 8} & 0_{8\times 8} & Q_1 & 0_{8\times 4} \\ 0_{4\times 8} & 0_{4\times 8} & 0_{4\times 8} & 0_{4\times 8} & 0_{4\times 8} & L_1 \end{bmatrix} \quad (6.230)$$

The components of the vector L are given by

$$\begin{aligned}
L_1 &= -\lambda_1^0 - a_1 R_1^* u_{q1}^s \lambda_2 - b_1 R_1^* x_3^s \lambda_2 - (a_2 u_{q2}^s + b_2 x_7^s) R_1 \lambda_6 \\
L_2 &= -\lambda_2^0 + K_{d1}^* \lambda_2 + x_1^s \lambda_4 - x_4^s \lambda_1 \\
L_3 &= -\lambda_3^0 - a_1 S_3^* u_{d1}^s \lambda_2 - C_1 S_3^* \lambda_3 + d_1 \lambda_3 - a_2 S_3 u_{d2}^s \lambda_6 - C_2 S_3 \lambda_7 - b_1 u_{d1}^s \lambda_2 \\
L_4 &= -\lambda_4^0 - a_1 S_1^* u_{d1}^s \lambda_2 - C_1 S_1^* \lambda_3 - a_2 S_1 u_{d2}^s \lambda_6 - C_2 S_1 \lambda_7 \\
L_5 &= -\lambda_5^0 - (a_1 u_{q1}^s + b_1 x_3^s) R_2^* \lambda_2 - (a_2 u_{q2}^s + b_2 x_7^s) R_2 \lambda_6 \qquad (6.231) \\
L_6 &= -\lambda_6^0 - x_8^s \lambda_5 + K_{d2}^* \lambda_6 + x_5^s \lambda_8 \\
L_7 &= -\lambda_7^0 - a_1 S_4^* u_{d1}^s \lambda_2 - C_1 S_4^* \lambda_3 - (a_2 S_4 + b_2) u_{d2}^s \lambda_6 + (d_2 - C_2 S_4) \lambda_7 \\
L_8 &= -\lambda_8^0 - a_1 S_2^* u_{d1}^s \lambda_2 - C_1 S_3^* \lambda_3 - a_2 S_2 u_{d2}^s \lambda_6 - C_2 S_2 \lambda_7 \\
L_{K+8} &= -\nu_K^0, \qquad K = 1, 2, \ldots, 32
\end{aligned}$$

The matrix R_{11} is of order 8×8, and its components are given by

$$\begin{gathered}
\gamma_{11} = k_1, \qquad \gamma_{12} = \tfrac{1}{2}\lambda_4 \\
\gamma_{13} = -\tfrac{1}{2}(a_1 R_1^* S_3^* \lambda_2 + a_2 R_1 S_3 \lambda_6 + b_1 R_1^* \lambda_2) \\
\gamma_{14} = -\tfrac{1}{2}(a_1 R_1^* S_1^* \lambda_2 + a_2 R_1 S_1 \lambda_6) \\
\gamma_{15} = 0, \qquad \gamma_{16} = 0 \\
\gamma_{17} = -\tfrac{1}{2}(a_1 R_1^* S_4^* \lambda_2 + a_2 R_1 S_r \lambda_6 + b_2 R_q \lambda_6) \\
\gamma_{18} = -\tfrac{1}{2}(a_1 R_1^* S_2^* \lambda_2 + a_2 R_1 S_2 \lambda_6) \\
\gamma_{22} = k_2, \qquad \gamma_{23} = 0 \\
\gamma_{24} = -\tfrac{1}{2}\lambda_1, \qquad \gamma_{25} = 0 \\
\gamma_{26} = 0, \qquad \gamma_{27} = 0 \\
\gamma_{28} = 0, \qquad \gamma_{33} = k_3 \\
\gamma_{34} = 0, \qquad \gamma_{35} = -\tfrac{1}{2}(a_1 S_3^* R_2^* \lambda_2 + b_1 R_2^* \lambda_2 + a_2 R_2 S_3 \lambda_6) \qquad (6.232)
\end{gathered}$$

$$\gamma_{36} = 0, \qquad \gamma_{37} = 0$$

$$\gamma_{38} = 0, \qquad \gamma_{44} = k_4$$

$$\gamma_{45} = -\tfrac{1}{2}(a_1 R_2^* S_1^* \lambda_2 + a_2 R_2 S_1 \lambda_6), \qquad \gamma_{46} = 0$$

$$\gamma_{47} = 0, \qquad \gamma_{48} = 0$$

$$\gamma_{55} = k_5, \qquad \gamma_{56} = \tfrac{1}{2}\lambda_8$$

$$\gamma_{57} = -\tfrac{1}{2}(a_1 R_2^* S_4^* \lambda_2 + a_2 S_4 R_2 \lambda_6 + b_2 R_2 \lambda_6)$$

$$\gamma_{58} = -\tfrac{1}{2}(a_1 R_2^* S_2^* \lambda_2 + a_2 R_2 S_2 \lambda_6)$$

$$\gamma_{66} = k_6, \qquad \gamma_{67} = 0$$

$$\gamma_{68} = -\tfrac{1}{2}\lambda_5, \qquad \gamma_{77} = k_7$$

$$\gamma_{78} = 0, \qquad \gamma_{88} = k_8$$

The matrix E_8 is a unit matrix of order 8×8. The matrices $0_{8\times4}$, $0_{8\times8}$, $0_{4\times8}$ are null matrices of order 8×4, 8×8, 4×8, respectively. Also,

$$M_1 = \begin{bmatrix} m_1 & & & & & & & \\ & m_2 & & & & 0 & & \\ & & m_3 & & & & & \\ & & & m_4 & & & & \\ & & & & m_5 & & & \\ & & & & & m_6 & & \\ & & & 0 & & & m_7 & \\ & & & & & & & m_8 \end{bmatrix} \tag{6.233}$$

$$N_1 = \begin{bmatrix} n_1 & & & & & & & \\ & n_2 & & & & 0 & & \\ & & n_3 & & & & & \\ & & & n_4 & & & & \\ & & & & n_5 & & & \\ & & & & & n_6 & & \\ & & & 0 & & & n_7 & \\ & & & & & & & n_8 \end{bmatrix} \tag{6.234}$$

$$P_1 = \begin{bmatrix} p_1 & & & & & & & \\ & p_2 & & & 0 & & & \\ & & p_3 & & & & & \\ & & & p_4 & & & & \\ & & & & p_5 & & & \\ & & & & & p_6 & & \\ & & & 0 & & & p_7 & \\ & & & & & & & p_8 \end{bmatrix} \tag{6.235}$$

$$Q_1 = \begin{bmatrix} q_1 & & & & & & & \\ & q_2 & & & & 0 & & \\ & & q_3 & & & & & \\ & & & q_4 & & & & \\ & & & & q_5 & & & \\ & & & & & q_6 & & \\ & & 0 & & & & q_7 & \\ & & & & & & & q_8 \end{bmatrix} \tag{6.236}$$

$$L_1 = \begin{bmatrix} l_1 & & 0 & \\ & l_2 & & \\ & & l_3 & \\ & 0 & & l_4 \end{bmatrix} \tag{6.237}$$

It will be noticed that the cost functional in equation (6.229) is composed of a boundary term and an integral part. To minimize J_1 in equation (6.229), one minimizes each term separately.

If one defines the following vector such that

$$V = R^{-1}L$$

then the integral part can be written in the following form by a process similar to completing the squares:

$$J_1 = \int_{t_0}^{t_f} \{[X + \tfrac{1}{2}V]^T L[X + \tfrac{1}{2}V] - \tfrac{1}{4}V^T RV\}\, dt + \text{boundary terms} \tag{6.238}$$

To minimize the integral part, one can consider only

$$J_{\text{in}} = \int_{t_0}^{t_f} (X + \tfrac{1}{2}V)^T L(X + \tfrac{1}{2}V)\, dt \tag{6.239}$$

because V is independent of X. Equation (6.239) defines a norm, hence equation (6.239) can be written as

$$\min_X J_{\text{in}} = \min_X[\|X + \tfrac{1}{2}V\|_L] \tag{6.240}$$

6.3.4. The Optimal Solution

The solution X which minimizes the cost functional J_{in} in equation (6.240) is given by

$$[X_\xi + \tfrac{1}{2}V] = [0] \tag{6.241}$$

Substituting from above into equation (6.421), one obtains the following optimality condition:

$$[L + 2RX_\xi] = [0] \tag{6.242}$$

Writing equation (6.242) explicitly, one obtains the following equations:

$$\begin{aligned}\dot{\lambda}_1 = &-R_1^*(a_1u_{q1}^s + b_1x_3^s)\lambda_2 - R_1(a_2u_{q2}^s + b_2x_7^s)\lambda_6\\ &+ 2k_1y_1 + \lambda_4y_2 - R_1^*(a_1S_3^* + b_1)\lambda_2y_3 - a_2R_1S_3\lambda_6y_3\\ &- a_1R_1^*S_1^*\lambda_2y_4 - a_2R_1S_1\lambda_6y_4 - a_1R_1^*S_4^*\lambda_2y_7 - a_1R_1^*S_4^*\lambda_2y_7\\ &- R_1(a_2S_4 + b_2)\lambda_6y_7 - a_1R_1^*S_2^*\lambda_2y_8 - a_2R_1S_2\lambda_6y_8\\ &- (\mu_1e_1 + \mu_2f_1 + \mu_3g_1 + \mu_5h_1)\end{aligned} \tag{6.243}$$

$$\begin{aligned}\dot{\lambda}_2 = &K_{d1}^*\lambda_2 + x_1^s\lambda_4 - x_4^s\lambda_1 + \lambda_4y_1 + 2k_2y_2 - \lambda_1y_4\\ &- (\mu_1e_2 + \mu_2f_2 + \mu_3g_2 + \mu_4h_2)\end{aligned} \tag{6.244}$$

$$\begin{aligned}\dot{\lambda}_3 = &-a_1S_3^*u_{d1}^s\lambda_2 - C_1S_3^*\lambda_3 + d_1^{\cdot}\lambda_3 - a_2S_3u_{d2}^s\lambda_6\\ &- C_2S_3\lambda_7 - R_1^*(a_1S_3^* + b_1)\lambda_2y_1 - a_2R_1S_3\lambda_6y_1\\ &+ 2k_3y_3 - R_2^*(a_1S_3^* + b_1)\lambda y_5 - a_2R_2S_3\lambda_6y_5\\ &- b_1u_{d1}^s\lambda_2 - (\mu_1e_3 + \mu_2f_3 + \mu_3g_3 + \mu_4h_3)\end{aligned} \tag{6.245}$$

$$\begin{aligned}\dot{\lambda}_4 = &-a_1S_1^*u_{d1}^s\lambda_2 - C_1S_1^*\lambda_3 - a_2S_1u_{d2}^s\lambda_6 - C_2S_1\lambda_7\\ &- a_1R_1^*S_1^*\lambda_2y_1 - a_2R_1S_1\lambda_6y_1 - \lambda_1y_2 + 2k_4y_4\\ &- a_1R_2^*S_1^*\lambda_2y_5 - a_2R_2S_1\lambda_6y_5\\ &- (\mu_1e_4 + \mu_2f_4 + \mu_3g_4 + \mu_4h_4)\end{aligned} \tag{6.246}$$

$$\dot{\lambda}_5 = -a_1 R_2^* u_{q1}^s \lambda_2 - b_1 R_2^* x_3^s \lambda_2 - R_2(a_2 u_{q2}^s + b_2 x_7^s)\lambda_6$$
$$- R_2^*(a_1 S_3^* + b_1)\lambda_2 y_3 - a_2 R_2 S_3 \lambda_6 y_3$$
$$- a_1 R_2^* S_1^* \lambda_2 y_4 - a_2 R_2 S_1 \lambda_6 y_4 + 2k_5 y_5 - a_1 R_2^* S_4^* \lambda_2 y_7$$
$$- R_2(a_2 S_4 + b_2)\lambda_6 y_7 - a_1 R_2^* S_2^* \lambda_2 y_8 - a_2 R_2 S_2 \lambda_6 y_8 + \lambda_8 y_6$$
$$- (\mu_1 e_5 + \mu_2 f_5 + \mu_3 g_5 + \mu_4 h_5) \tag{6.247}$$

$$\dot{\lambda}_6 = -x_8^s \lambda_5 + K_{d2}^* \lambda_6 + x_5^s \lambda_8 + \lambda_8 y_5 + 2k_6 y_6$$
$$- \lambda_5 y_8 - (\mu_1 e_6 + \mu_2 f_6 + \mu_3 g_6 + \mu_4 h_6) \tag{6.248}$$

$$\dot{\lambda}_7 = -a_1 S_4^* u_{d1}^s \lambda_2 - C_1 S_4^* \lambda_3 - a_2 S_4 u_{d2}^s \lambda_6$$
$$+ d_2 \lambda_7 - C_2 S_4 \lambda_7 - a_1 R_1^* S_4^* \lambda_2 y_1 - (a_2 S_4 + b_2)\lambda_6 y_1$$
$$- a_1 R_2^* S_4^* \lambda_2 y_5 - R_2(a_2 S_4 + b_2)\lambda_6 y_5 + 2k_7 y_7$$
$$- b_2 u_{d2}^s \lambda_6 - (\mu_1 e_7 + \mu_2 f_7 + \mu_3 g_7 + \mu_4 h_7) \tag{6.249}$$

$$\dot{\lambda}_8 = -a_1 S_2^* u_{d1}^s \lambda_2 - C_1 S_2^* \lambda_3 - a_2 S_2 u_{d2}^s \lambda_6$$
$$- C_2 S_2 \lambda_7 - a_1 R_1^* S_2^* \lambda_2 y_1 - a_2 R_1 S_2 \lambda_6 y_1$$
$$- a_1 R_2^* S_2^* \lambda_2 y_5 - a_2 R_2 S_2 \lambda_6 y_5 - \lambda_5 y_6 + 2k_8 y_8$$
$$- (\mu_1 e_8 + \mu_2 f_8 + \mu_3 g_8 + \mu_4 h_8) \tag{6.250}$$

and

$$\dot{\nu}_1 = -\mu_1 y_1 + 2m_1 e_1, \qquad \dot{\nu}_2 = -\mu_1 y_2 + 2m_2 e_2 \tag{6.251}$$
$$\dot{\nu}_3 = -\mu_1 y_3 + 2m_3 e_3, \qquad \dot{\nu}_4 = -\mu_1 y_4 + 2m_4 e_4 \tag{6.252}$$
$$\dot{\nu}_5 = -\mu_1 y_5 + 2m_5 e_5, \qquad \dot{\nu} = -\mu_1 y_6 + 2m_6 e_6 \tag{6.253}$$
$$\dot{\nu}_7 = -\mu_1 y_7 + 2m_7 e_7, \qquad \dot{\nu}_8 = -\mu_1 y_8 + 2m_8 e_8 \tag{6.254}$$
$$\dot{\nu}_9 = -\mu_2 y_1 + 2n_1 f_1, \qquad \dot{\nu}_{10} = -\mu_2 y_2 + 2n_2 f_2 \tag{6.255}$$
$$\dot{\nu}_{11} = -\mu_2 y_3 + 2n_3 f_3, \qquad \dot{\nu}_{12} = -\mu_2 y_4 + 2n_4 f_4 \tag{6.256}$$
$$\dot{\nu}_{13} = -\mu_2 y_5 + 2n_5 f_5, \qquad \dot{\nu}_{14} = -\mu_2 y_6 + 2n_6 f_6 \tag{6.257}$$
$$\dot{\nu}_{15} = -\mu_2 y_7 + 2n_7 f_7, \qquad \dot{\nu}_{16} = -\mu_2 y_8 + 2n_8 f_8 \tag{6.258}$$
$$\dot{\nu}_{17} = -\mu_3 y_1 + 2p_1 g_1, \qquad \dot{\nu}_{18} = -\mu_3 y_2 + 2p_2 g_2 \tag{6.259}$$
$$\dot{\nu}_{19} = -\mu_3 y_3 + 2p_3 g_3, \qquad \dot{\nu}_{20} = -\mu_3 y_4 + 2p_4 g_4 \tag{6.260}$$
$$\dot{\nu}_{21} = -\mu_3 y_5 + 2p_5 g_5, \qquad \dot{\nu}_{22} = -\mu_3 y_6 + 2p_6 g_6 \tag{6.261}$$
$$\dot{\nu}_{23} = -\mu_3 y_7 + 2p_7 g_7, \qquad \dot{\nu}_{24} = -\mu_3 y_8 + 2p_8 g_8 \tag{6.262}$$

$$\dot{\nu}_{25} = -\mu_4 y_1 + 2q_1 h_1, \qquad \dot{\nu}_{26} = -\mu_4 y_2 + 2q_2 h_2 \tag{6.263}$$

$$\dot{\nu}_{27} = -\mu_4 y_3 + 2q_3 h_3, \qquad \dot{\nu}_{28} = -\mu_4 y_4 + 2q_4 h_4 \tag{6.264}$$

$$\dot{\nu}_{29} = -\mu_4 y_5 + 2q_5 h_5, \qquad \dot{\nu}_{30} = -\mu_4 y_6 + 2q_6 h_6 \tag{6.265}$$

$$\dot{\nu}_{31} = -\mu_4 y_7 + 2q_7 h_7, \qquad \dot{\nu}_{32} = -\mu_4 y_8 + 2q_8 h_8 \tag{6.266}$$

and

$$\mu_1 = (1/M_1)\lambda_2 - 2l_1 u_1, \qquad \mu_2 = \lambda_3 - 2l_2 v_1 \tag{6.267}$$

$$\mu_3 = (1/M_2)\lambda_6 - 2l_3 u_2, \qquad \mu_4 = \lambda_7 - 2l_4 v_2 \tag{6.268}$$

together with equations (6.207)-(6.214), (6.220), (6.221), and the boundary conditions

$$y_i(t_0) = 0, \qquad \lambda_i(t_f) = 0, \qquad i = 1, 2, \ldots, 8 \tag{6.269}$$

$$\nu_j(t_0) = 0, \qquad \nu_i(t_f) = 0, \qquad j = 1, 2, \ldots, 32 \tag{6.270}$$

The nonlinear optimal control problem with linear feedback has been reduced to the two-point boundary value problem which will be numerically solved in the following section.

Integrating both sides of equations (6.251)-(6.266) from t_0 to t_f, and using the boundary conditions (6.267) and (6.268), one obtains for $i = 1, 2, \ldots, 8$

$$\int_{t_0}^{t_f} (-\mu_1 y_i + 2m_i e_i)\, dt = 0 \tag{6.271}$$

$$\int_{t_0}^{t_f} (-\mu_2 y_i + 2n_i f_i)\, dt = 0 \tag{6.272}$$

$$\int_{t_0}^{t_f} (-\mu_3 y_i + 2p_i g_i)\, dt = 0 \tag{6.273}$$

$$\int_{t_0}^{t_f} (-\mu_4 y_i + 2q_i p_i)\, dt = 0 \tag{6.274}$$

These relations will be used in Section 6.3.5.

For the special case of local control, the augmented cost functional is given by

$$J_1 = \int_{t_0}^{t_f} F_L\, dt \tag{6.275}$$

where

$$
\begin{aligned}
F_L = {} & \sum_{i=1}^{8} k_i y_i^2 + \sum_{i=1}^{4} (m_i e_i^2 + n_i f_i^2 + p_i g_i^2 + q_i h_i^2) \\
& + l_1 u_1^2 + l_2 v_1^2 + l_3 u_2^2 + l_4 v_2^2 \\
& + \mu_1 \left(u_1 - \sum_{i=1}^{4} e_i y_i \right) + \mu_2 \left(v_1 - \sum_{i=1}^{4} f_i y_i \right) \\
& + \mu_3 \left(u_2 - \sum_{i=1}^{4} g_i y_{i+4} \right) + \mu_4 \left(v_2 - \sum_{i=1}^{4} h_i y_{i+4} \right) \\
& + \sum_{i=1}^{4} \nu_i (de_i/dt) + \sum_{i=1}^{4} \nu_{i+4} (df_i/dt) \\
& + \sum_{i=1}^{4} \nu_{i+8} (dg_i/dt) + \sum_{i+4}^{4} \nu_{i+12} (dh_i/dt) \\
& + \lambda_1 (dy_1/dt - x_4^s y_2 - y_2 y_4) \\
& + \lambda_2 \{ dy_2/dt + a_1 (u_{d1}^s u_{q1}^y + u_{q1}^s u_{d1}^y + u_{d1}^y u_{q1}^y) \\
& - b_1 (x_3^s u_{d1}^y + u_{d1}^s y_3 + u_{d1}^y y_3) - (1/M_1) u_1 \\
& + K_{d1}^* y_2 - (1/M_1) K [u(t - t_0) - u(t - \tau)] \} \\
& + \lambda_3 (dy_3/dt - v_1 - C_1 u_{q1}^y + d_1 y_3) \\
& + \lambda_4 (dy_4/dt + x_1^s y_2 + y_1 y_2) \\
& + \lambda_5 (dy_5/dt - x_8^s y_6 - y_6 y_8) \\
& + \lambda_6 [dy_6/dt - a_2 (u_{d2}^s u_{q2}^y + u_{q2}^s u_{d2}^y + u_{d2}^y u_{q2}^y) \\
& - b_2 (x_7^s u_{d2}^y + u_{d2}^s y_7 + u_{d2}^y y_7) - (1/M_2) u_2 + K_{d2}^* y_6] \\
& + \lambda_7 (dy_7/dt - \nu_2 - c_2 u_{q2}^y + d_2 y_7) \\
& - \lambda_8 (dy_8/dt + x_5^s y_6 + y_5 y_6)
\end{aligned}
\tag{6.276}
$$

Note that the number of Lagrange multipliers has been reduced from 44, as in the case of global control, to 28, as in this case. The procedures used for formulating the optimality equations for global control can also be used in this case.

The optimal solutions for the case of local control satisfy the following relations:

$$\dot{\lambda}_1 = -R_1^*(a_1 u_{q1}^s + b_1 x_3^s)\lambda_2 - R_1(a_2 u_{q2}^s + b_2 x_7^s)\lambda_6$$
$$+ 2k_1 y_1 + \lambda_4 y_2 - R_1^*(a_1 S_3^* + b_1)\lambda_2 y_3 - a_2 R_1 S_3 \lambda_6 y_3$$
$$- a_1 R_1^* S_1^* \lambda_2 y_4 - a_2 R_1 S_1 \lambda_6 y_4 - a_1 R_1^* S_4^* \lambda_2 y_7 - R_1(a_2 S_4 + b_2)\lambda_6 y_7$$
$$- a_1 R_1^* S_2^* \lambda_2 y_8 - a_2 R_1 S_2 \lambda_6 y_8 - (\mu_1 e_1 + \mu_2 f_1) \tag{6.277}$$

$$\dot{\lambda}_2 = K_{d1}^* \lambda_2 + x_1^s \lambda_4 - x_4^s \lambda_1 + \lambda_4 y_1 + 2k_2 y_2 - \lambda_1 y_4 - (\mu_1 e_2 + \mu_2 f_2) \tag{6.278}$$

$$\dot{\lambda}_3 = -a_1 S_3^* u_{d1}^s \lambda_2 - C_1 S_3^* \lambda_3 + d_1 \lambda_3 \quad a_2 S_3 u_{d2}^s \lambda_6 - C_2 S_3 \lambda_7$$
$$- R_1^*(a_1 S_3^* + b_1)\lambda_2 y_1 - a_2 R_1 S_3 \lambda_6 y_1 + 2k_3 y_3$$
$$- R_2^*(a_1 S_3^* + b_1)\lambda_2 y_5 - a_2 R_2 S_3 \lambda_6 y_5 - b_1 u_{d1}^s \lambda_2 - (\mu_1 e_3 + \mu_2 f_3) \tag{6.279}$$

$$\dot{\lambda}_4 = -a_1 S_1^* u_{d1}^s \lambda_2 - C_1 S_1^* \lambda_3 - a_2 S_1 u_{d2}^s \lambda_6 - C_2 S_1 \lambda_7$$
$$- a_1 R_1^* S_1^* \lambda_2 y_1 - a_2 R_1 S_1 \lambda_6 y_1 - \lambda_1 y_2 + 2k_4 y_4$$
$$- a_1 R_2^* S_1^* \lambda_2 y_5 - a_2 R_2 S_1 \lambda_6 y_5 - (\mu_1 e_4 + \mu_2 f_4) \tag{6.280}$$

$$\dot{\lambda}_5 = -a_1 R_2^* u_{q1}^s \lambda_2 - b_1 R_2^* x_3^s \lambda_2 - R_2(a_2 u_{q2}^s + b_2 x_7^s)\lambda_6$$
$$- R_2^*(a_1 S_3^* + b_1)\lambda_2 y_3 - a_2 R_2 S_e \lambda_6 y_3 - a_1 R_2^* S_1^* \lambda_2 y_4$$
$$- a_2 R_2 S_1 \lambda_6 y_4 + 2k_5 y_5 - a_1 R_2^* S_4^* \lambda_2 y_7 - R_2(a_2 S_4 + b_2)\lambda_6 y_7$$
$$- a_1 R_2^* S_2^* \lambda_2 y_8 - a_2 R_2 S_2 \lambda_6 y_8 + \lambda_8 y_6 - (\mu_3 g_1 + \mu_4 h_1) \tag{6.281}$$

$$\dot{\lambda}_6 = -x_8^s \lambda_5 + K_{d2}^* \lambda_6 + x_5^s \lambda_8 + \lambda_8 y_5 + 2k_6 y_6 - \lambda_5 y_8 - (\mu_3 g_2 + \mu_4 h_2) \tag{6.282}$$

$$\dot{\lambda}_7 = -\, a_1 S_4^* u_{d1}^s \lambda_2 - C_1 S_4^* \lambda_3 - a_2 S_4 u_{d2}^s \lambda_6 + d_2 \lambda_7$$
$$- C_2 S_4 \lambda_7 - a_1 R_1^* S_4^* \lambda_2 y_1 - (a_2 S_4 + b_2) R_1 \lambda_6 y_1$$
$$- a_1 R_2^* S_4^* \lambda_2 y_5 - R_2(a_2 S_4 + b_2)\lambda_6 y_5 + 2k_7 y_7$$
$$- b_2 u_{d2}^s \lambda_6 - (\mu_3 g_3 + \mu_4 h_3) \tag{6.283}$$

$$\dot{\lambda}_8 = -a_1 S_2^* u_{d1}^s \lambda_2 - C_1 S_2^* \lambda_3 - a_2 S_2 u_{d2}^s \lambda_6 - C_2 S_2 \lambda_7$$
$$- a_1 R_1^* S_2^* \lambda_2 y_1 - a_2 R_1 S_2 \lambda_6 y_1 - a_1 R_2^* S_2^* \lambda_2 y_5$$
$$- a_2 R_2 S_2 \lambda_6 y_5 - \lambda_5 y_6 + 2k_8 y_8 - (\mu_3 g_4 + \mu_4 h_4) \tag{6.284}$$

and

$$\mathring{\nu}_1 = -\mu_1 y_1 + 2m_1 e_1, \qquad \mathring{\nu}_2 = -\mu_1 y_2 + 2m_2 e_2 \tag{6.285}$$

$$\mathring{\nu}_3 = -\mu_1 y_3 + 2m_3 e_3, \qquad \mathring{\nu}_4 = -\mu_1 y_4 + 2m_4 e_4 \tag{6.286}$$

$$\mathring{\nu}_5 = -\mu_2 y_1 + 2n_1 f_1, \qquad \mathring{\nu}_6 = -\mu_2 y_2 + 2n_2 f_2 \tag{6.287}$$

$$\mathring{\nu}_7 = -\mu_2 y_3 + 2n_3 f_e, \qquad \mathring{\nu}_8 = -\mu_2 y_4 + 2n_4 f_4 \tag{6.288}$$

$$\mathring{\nu}_9 = -\mu_3 y_5 + 2p_1 g_1, \qquad \mathring{\nu}_{10} = -\mu_3 y_6 + 2p_2 g_2 \tag{6.289}$$

$$\mathring{\nu}_{11} = -\mu_3 y_7 + 2p_3 g_3, \qquad \mathring{\nu}_{12} = -\mu_3 y_8 + 2p_4 g_4 \tag{6.290}$$

$$\mathring{\nu}_{13} = -\mu_4 y_5 + 2q_1 h_1, \qquad \mathring{\nu}_{14} = -\mu_4 y_6 + 2q_2 h_2 \tag{6.291}$$

$$\mathring{\nu}_{15} = -\mu_4 y_7 + 2q_3 h_3, \qquad \mathring{\nu}_{16} = -\mu_4 y_8 + 2q_4 h_4 \tag{6.292}$$

and

$$\mu_1 = (1/M_1)\lambda_2 - 2l_1 u_1, \qquad \mu_2 = \lambda_3 - 2l_2 v_1 \tag{6.293}$$

$$\mu_3 = (1/M_2)\lambda_6 - 2l_3 u_2, \qquad \mu_4 = \lambda_7 - 2l_4 v_2 \tag{6.294}$$

6.3.5. Computation Techniques

The problem being solved can further be reduced to one of parameter optimization. Substituting equations (6.220) and (6.221) into equations (6.207)-(6.214) we obtain a system of differential equations for $y_1, y_2, \ldots, y_8$ after the parameters $e_1, e_2, \ldots, e_g, f_1, f_2, \ldots, f_8, g_1, g_2, \ldots, g_8, h_1, h_2, \ldots, h_8$ have been determined. Let $\overline{y_1}(t), \overline{y_2}(t), \ldots, \overline{y_8}(t)$ be the solutions of the system of differential equations with initial conditions

$$y_i(t_0) = 0, \qquad i = 1, 2, \ldots, 8 \tag{6.295}$$

Substituting $\{\bar{y}_i(t),\ i = 1, 2, \ldots, 8\}$ into equations (6.267) and (6.268), one has

$$\mu_1 = (1/M_1)\lambda_2 - 2l_1 \sum_{i=1}^{8} e_i \bar{y}_i(t) \tag{6.296}$$

$$\mu_2 = \lambda_3 - 2l_2 \sum_{i=1}^{8} f_i \bar{y}_i(t) \tag{6.297}$$

$$\mu_3 = (1/M_2)\lambda_6 - 2l_3 \sum_{i=1}^{8} g_i \bar{y}_i(t) \tag{6.298}$$

$$\mu_4 = \lambda_7 - 2l_4 \sum_{i=1}^{8} h_i \bar{y}_i(t) \tag{6.299}$$

Substituting equations (6.296)-(6.299) and $\{\bar{y}_i(t),\ i = 1, 2, \ldots, 8\}$ into equations (6.243)-(6.250), a system of differential equations for λ_1,

$\lambda_2, \ldots, \lambda_8$ is obtained. This system with terminal condition

$$\lambda_i(t_f) = 0, \qquad i = 1, 2, \ldots, 8 \tag{6.300}$$

can be solved by integrating backward from t_f to t_0. First, let the solutions to the system of differential equations be $\bar{\lambda}_1(t), \bar{\lambda}_2(t), \ldots, \bar{\lambda}_8(t)$. Then, the problem is reduced to selecting 32 parameters e_i, f_i, g_i, h_i, $i = 1, 2, \ldots, 8$, such that

$$\int_{t_0}^{t_f} \begin{bmatrix} -\bar{u}_1(t)\bar{y}_i(t) + 2m_i e_i \\ -\bar{\mu}_2(t)\bar{y}_i(t) + 2n_i f_i \\ -\bar{\mu}_3(t)\bar{y}_i(t) + 2p_i g_i \\ -\bar{\mu}_4(t)\bar{y}_i(t) + 2q_i h_i \end{bmatrix} dt = 0, \qquad i = 1, 2, \ldots, 8 \tag{6.301}$$

and the cost functional

$$\begin{aligned} J = \int_{t_0}^{t_f} \Bigg\{ & \left[\sum_{i=1}^{8} k_i \bar{y}_i(t)^2 + m_i e_i^2 + n_i f_i^2 + p_i g_i^2 + q_i h_i^2 \right] \\ & + l_1 \left[\sum_{i=1}^{8} e_i \bar{y}_i(t) \right]^2 + l_2 \left[\sum_{i=1}^{8} f_i \bar{y}_i(t) \right]^2 \\ & + l_3 \left[\sum_{i=1}^{8} g_i \bar{y}_i(t) \right]^2 + l_4 \left[\sum_{i=1}^{8} h_i \bar{y}_i(t) \right]^2 \Bigg\} dt \end{aligned} \tag{6.302}$$

is minimized, where

$$\bar{\mu}_1(t) = (1/M_1)\bar{\lambda}_2(t) - 2l_1 \sum_{i=1}^{8} e_i \bar{y}_i(t) \tag{6.303}$$

$$\bar{\mu}_2(t) = \lambda_3(t) - 2l_2 \sum_{i=1}^{8} f_i \bar{y}_i(t) \tag{6.304}$$

$$\bar{\mu}_3(t) = (1/M_2)\lambda_6(t) - 2l_3 \sum_{i=1}^{8} g_i \bar{y}_i(t) \tag{6.305}$$

$$\bar{\mu}_4(t) = \bar{\lambda}_7(t) - 2l_4 \sum_{i=1}^{8} h_i \bar{y}_i(t) \tag{6.306}$$

This problem can be solved numerically by the gradient method. For $i = 1, 2, \ldots, 8$, let

$$a_i(t) = \partial H/\partial e_i = -\mu_1(t)y_i(t) + 2m_i e_i \tag{6.307}$$

$$b_i(t) = \partial H/\partial f_i = -\mu_2(t)y_i(t) + 2n_i f_i \tag{6.308}$$

$$c_i(t) = \partial H/\partial g_i = -\mu_3(t)y_i(t) + 2p_i g_i \tag{6.309}$$

$$d_i(t) = \partial H/\partial h_i = -\mu_4(t)y_i(t) + 2q_i h_i \tag{6.310}$$

and

$$a_i = \int_{t_0}^{t_f} a_i(t)\,dt, \qquad b_i = \int_{t_0}^{t_f} b_i(t)\,dt \tag{6.311}$$

$$c_i = \int_{t_0}^{t_f} c_i(t)\,dt, \qquad d_i = \int_{t_0}^{t_f} d_i(t)\,dt \tag{6.312}$$

where

$$\begin{aligned}
\mu_1(t) &= (1/M_1)\lambda_2(t) - 2l_1 \sum_{i=1}^{8} e_i y_i(t) \\
\mu_2(t) &= \lambda_3(t) - 2l_2 \sum_{i=1}^{8} f_i y_i(t) \\
\mu_3(t) &= (1/M_2)\lambda_6(t) - 2l_3 \sum_{i=1}^{8} g_i y_i(t) \\
\mu_4(t) &= \lambda_7(t) - 2l_4 \sum_{i=1}^{8} h_i y_i(t)
\end{aligned} \tag{6.313}$$

and $\{y_i(t),\ i = 1, 2, \ldots, 8\}$ and $\{\lambda_i(t),\ i = 1, 2, \ldots, 8\}$ are the solutions of the system of equations (6.207)-(6.214) and (6.243)-(6.250) for a preselected set of parameters e_i, f_i, g_i, h_i, $i = 1, 2, \ldots, 8$.

The iteration procedure is given by deduction. A first guess of the control parameters $\{e_i^0, f_i^0, g_i^0, h_i^0, i = 1, 2, \ldots, 8\}$ is arbitrarily assigned. If the control parameters of the jth iteration $\{e_i^j, f_i^j, g_i^j, h_i^j, i = 1, 2, \ldots, 8\}$ are determined, then the ones of the $(j+1)$th iteration $\{e_i^{j+1}, f_i^{j+1}, g_i^{j+1}, h_i^{j+1}, i = 1, 2, \ldots, 8\}$ take the form for $i = 1, 2, \ldots, 8$

$$e_i^{j+1} = e_i^j - \alpha_i^j a_i^j \tag{6.314}$$

$$f_i^{i+1} = f_i^j - \beta_i^j b_i^j \tag{6.315}$$

$$g_i^{j+1} = g_i^j - \gamma_i^j c_i^j \tag{6.316}$$

$$h_i^{j+1} = h_i^j - \delta_i^j d_i^j \tag{6.317}$$

where the α_i^j's, β_i^j's, γ_i^j's, δ_i^j's are suitable selected constants. The a_i^j's, b_i^j's, c_i^j's, d_i^j's, which also depend on the iteration number j, are given by equations (6.307)-(6.317), by using the control parameters of the jth iteration.

For all i, $i = 1, 2, \ldots, 8$, the stopping criteria are given by

$$|a_i| < \varepsilon \tag{6.318}$$

$$|b_i| < \varepsilon \tag{6.319}$$

$$|c_i| < \varepsilon \tag{6.320}$$

$$|d_i| < \varepsilon \tag{6.321}$$

and the cost functional is to be kept reducing with increasing j. Condition (6.321) is selected as one of the stopping criteria because, numerically speaking, (6.321) is equivalent to (6.303)-(6.306) if ε is a small positive number.

In practical computation, the α_i^j's, β_i^j's, γ_i^j's, δ_i^j's are taken to be equal to, say, ω^j, which is used to obtain control parameters of the $(j+1)$th iteration from the ones of the jth iteration. It is noted that ω^j varies with the number of iterations j for the fixed strength of torque pulse K.

ω^j is determined as follows. Suppose that the control parameters of the jth iteration, for which the stopping criteria are not met, are obtained. The a_i^j's, b_i^j's, c_i^j's, d_i^j's are obtained from equations (6.307)-(6.317) by using the control parameters of the jth iteration. For given $\bar{\omega}$, letting (for $i = 1, 2, \ldots, 8$)

$$\alpha_i^j = \beta_i^j = \gamma_i^j = \delta_i^j = \bar{\omega}$$

one obtains

$$\begin{aligned} \bar{e}_i &= e_i - \bar{\omega} a_i^j \\ \bar{f}_i &= f_i - \bar{\omega} b_i^j \\ \bar{g}_i &= g_i - \bar{\omega} c_i^j \\ \bar{h}_i &= h_i - \bar{\omega} d_i^j \end{aligned} \tag{6.322}$$

Substituting

$$u_1 = \sum_{i=1}^{8} \bar{e}_i y_i, \qquad v_1 = \sum_{i=1}^{8} \bar{f}_i y_i$$

$$u_2 = \sum_{i=1}^{8} \bar{g}_i y_i, \qquad v_2 = \sum_{i=1}^{8} \bar{h}_i y_i$$

into equations (6.207)-(6.204) with the help of the initial conditions (6.269)-(6.270) the solutions for the differential equations (6.207)-(6.214) are obtained. By further substitution of these solutions into equations (6.220)-(6.221) and equation (6.225), the cost functional $\bar{J}$ can be obtained by integration. At this point, it is said that $\bar{J}$ is a function of $\bar{\omega}$. ω^j is selected such that $\bar{J}$ is minimized. In so doing, the cost functional can always be reduced.

The differential equations (6.207)-(6.214) and (6.243)-(6.250) are numerically solved forward and backward, respectively, by fourth-order Runge-Kutta algorithms with step size $h = 0.005$. This step size is good enough for practical cases ($K = 0.0$ to $K = 0.50$); but, for a large disturbance (such as $K = 2.0$), a smaller step size is required to produce accurate results.

The optimal solutions are obtained numerically from $K = 0.0$ to $K = 0.50$ for both global and local controls. For testing purposes, the optimal solutions for two separated machine systems (in which case the coupling between the two machines is broken) are also obtained.

The optimal parameters for the case $K = 0.0$, all of which are zeros, are first obtained. These values are then used as a first guess to find the optimal parameters for the case $K = 0.05$. For succeeding search for the optimal parameters ($K \geq 0.10$), the same procedures are applied. This method usually provides a good initial guess, because of the continuity of the optimal parameters with respect to K; it has been observed that the rate of convergence is fast.

The results shown in Figs. 6.4–6.15 were obtained with $x_3 = 0.10$ p.u. It will be noted that the deviations from the steady state for all state variables y_i, $i = 1, \ldots, 8$, have been recorded. The torque pulse is applied to generator number 1, and the state variable deviations y_1, y_2, y_3, y_4 show the corresponding variations in Figs. 6.4–6.7. It will be observed that the corresponding deviations in state variable deviations y_5, y_6, y_7, y_8 for the second generator are of much smaller magnitude, as is indicated in Figs. 6.8–6.11. This was also found to be the case for smaller values of the coupling impedance x_3 than that used above. To illustrate this, Tables 6.3–6.6 have been included. This leads one to conclude that, for the case at hand, local control [see equations (6.223) and (6.224)] is adequate.

Figures 6.12 and 6.13 give the optimal feedback parameters e_i ($i = 1$–4) and f_i ($i = 1$–4) [see equations (6.220) and (6.221)] as a function of the strength K of the torque pulse for machine number 1. Figures 6.14 and 6.15 give the corresponding values for machine number 2. Thus, essentially an adaptive scheme has been developed. The feedback state variables are rather easily obtained from measurement of frequency and terminal

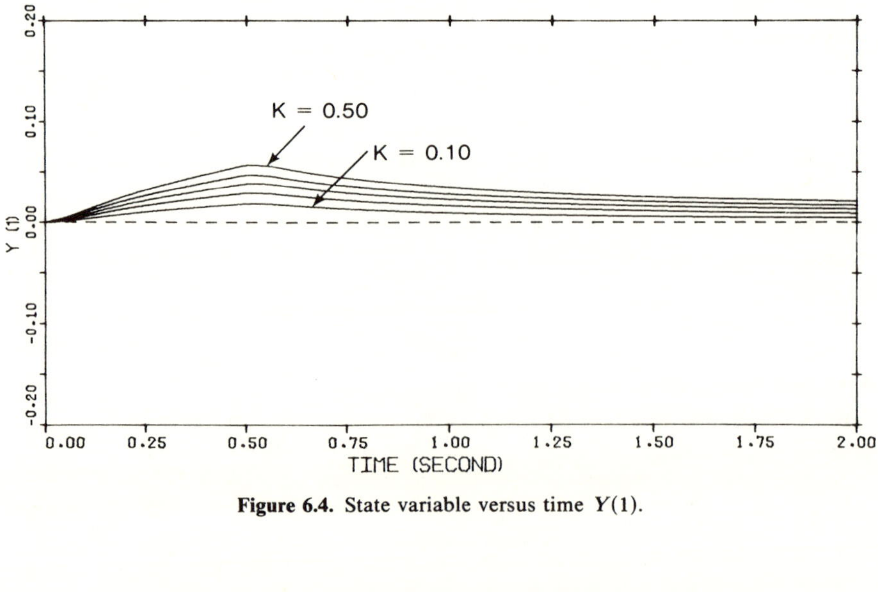

Figure 6.4. State variable versus time $Y(1)$.

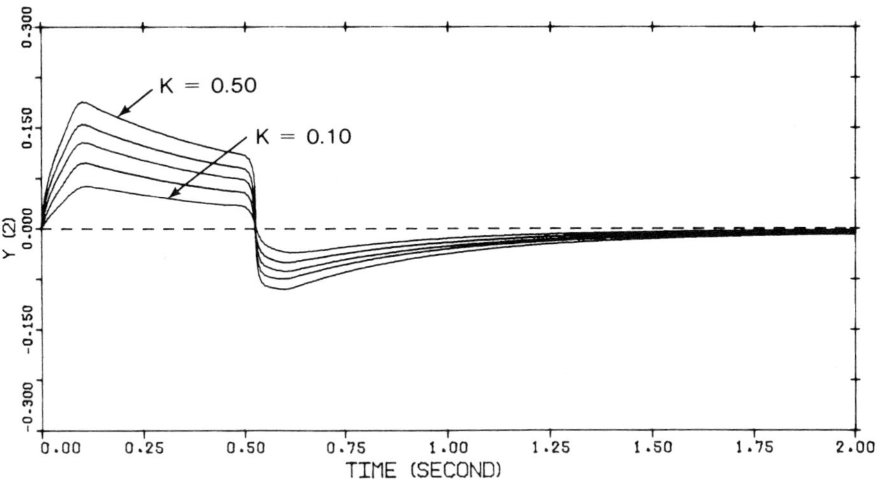

Figure 6.5. State variable versus time $Y(2)$.

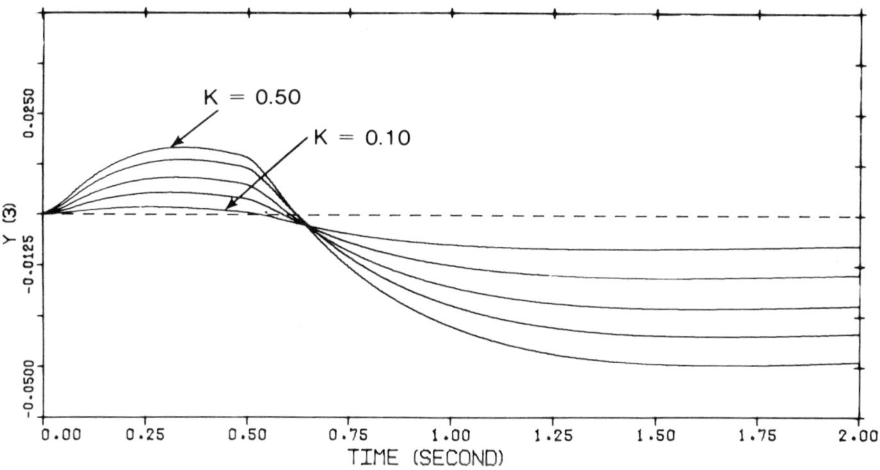

Figure 6.6. State variable versus time $Y(3)$.

voltage. This voltage is related to y_3 by a quadratic expression for machine number one.

6.4. Realistic Feedback Control of Two Interconnected Turbogenerators

In this section, a general nonlinear model of a system with two identical interconnected turbogenerators, which are connected to an infinite bus, is

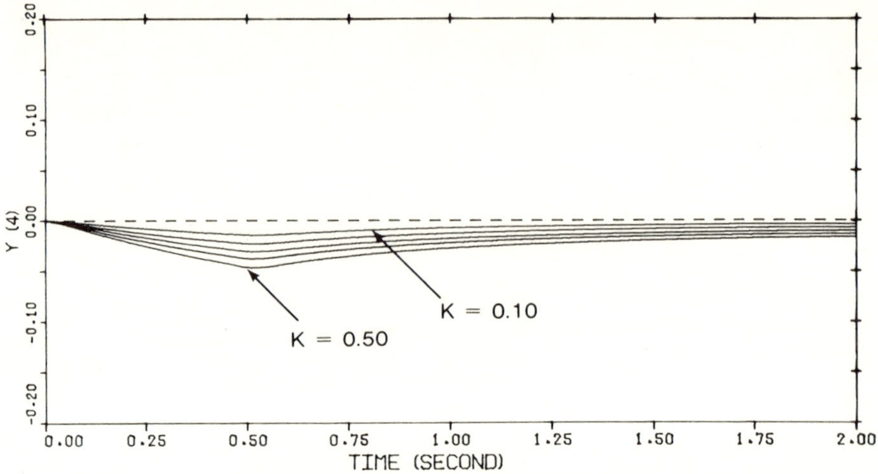

Figure 6.7. State variable versus time $Y(4)$.

considered. The control of the system is effected through the control of field voltages and turbine torques. One of the machines is disturbed from the steady state by a torque pulse applied directly to the shaft. While such a disturbance is not realistic, it gives valuable information about the system performance.

In this instance, it is proposed to include the transfer functions of the governors, the turbines, and the voltage regulators. Inclusion of these transfer functions provides a more realistic model of the machine and its control system; however, one does increase the dimensions of the system.

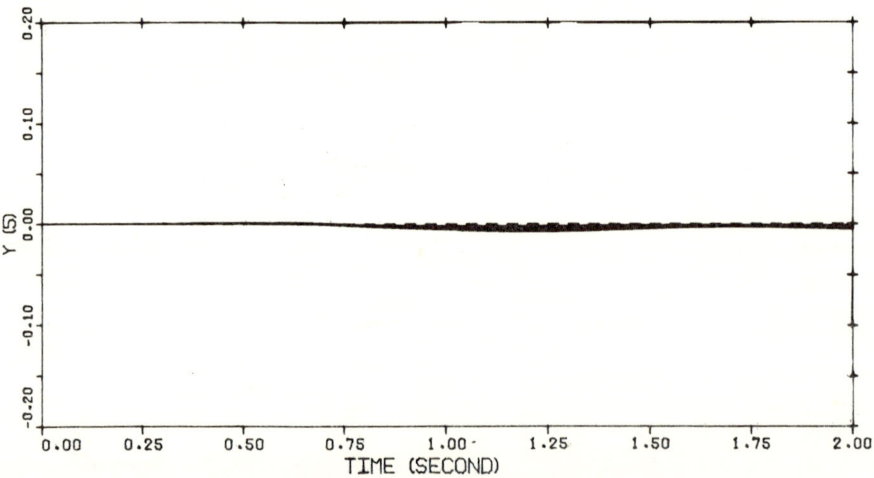

Figure 6.8. State variable versus time $Y(5)$.

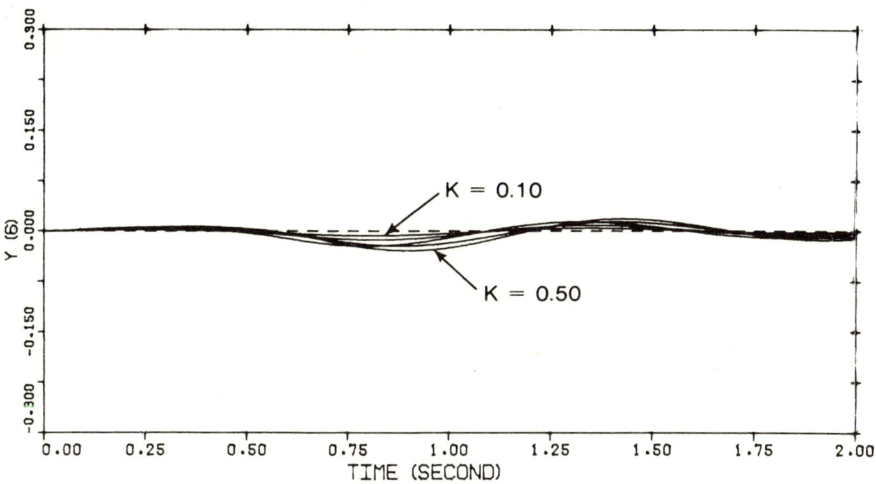

Figure 6.9. State variable versus time $Y(6)$.

The controls are modeled as linear combinations of the state variables, but the machine equations are not linearized. Therefore, the model parameters, which are assumed to be time-independent, generally depend on the strength and duration of the pulse, contrary to the linearized machine models where the equations themselves are linearized. Here, a different set of feedback parameters is needed for each different pulse strength. This linear feedback model is considered, because it leads to a simple implementation, as compared with the work mentioned in Section 6.2.

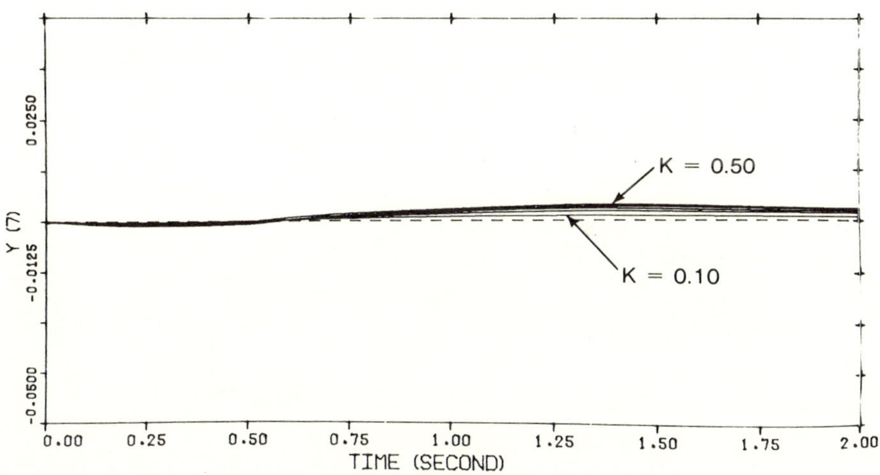

Figure 6.10. State variable versus time $Y(7)$.

Table 6.3. State Space Variables, $K = 0.50$ and $x_3 = 0.10$

t	$Y(1)$	$Y(2)$	$Y(3)$	$Y(4)$	$Y(5)$	$Y(6)$	$Y(7)$	$Y(8)$
0.00	0.00000000	0.00000000	0.00000000	0.00000000	0.00000000	0.00000000	0.00000000	0.00000000
0.05	0.00484201	0.18549424	0.00335879	−0.00367985	0.00000116	0.00014539	0.00001841	−0.00000088
0.10	0.01231036	0.18613005	0.00798628	−0.00944646	0.00001729	0.00080799	0.00011188	−0.00001307
0.15	0.01940334	0.17340070	0.01166163	−0.01502873	0.00006664	0.00179007	0.00027985	−0.00005037
0.20	0.02594646	0.16091204	0.01433023	−0.02027147	0.00015603	0.00276143	0.00050732	−0.00011795
0.25	0.03198352	0.14963216	0.01609844	−0.02518969	0.00027946	0.00324171	0.00078002	−0.00021129
0.30	0.03756937	0.13952351	0.01707387	−0.02981089	0.00042006	0.00353641	0.00108589	−0.00031764
0.35	0.04275408	0.13047624	0.01735344	−0.03416200	0.00055308	0.00295554	0.00141532	−0.00041830
0.40	0.04758217	0.12238312	0.01702349	−0.03826823	0.00064945	0.00162253	0.00176125	−0.00049126
0.45	0.05209312	0.11514580	0.01616060	−0.04215291	0.00067959	−0.00042599	0.00211904	−0.00051407
0.50	0.05632190	0.10867447	0.01483258	−0.04583751	0.00061687	−0.00306933	0.00248618	−0.00046659
0.60	0.05243769	−0.08737367	0.00308465	−0.04244777	0.00011540	−0.01045243	0.00312958	−0.00008721
0.70	0.04643534	−0.07012326	−0.00810856	−0.03727909	−0.00105835	−0.01921137	0.00352302	0.00079888
0.80	0.04163221	−0.05576731	−0.01657268	−0.03320250	−0.02283335	−0.02458637	0.00379021	0.00213377
0.90	0.03778173	−0.04466845	−0.02293055	−0.02997201	−0.00481140	−0.02363166	0.00402644	0.00361414
1.00	0.03467484	−0.03606202	−0.02766723	−0.02738940	−0.00646341	−0.01628089	0.00426570	0.00484464
1.10	0.03214885	−0.02937532	−0.03114916	−0.02530533	−0.00735575	−0.00504377	0.00448943	0.00550706
1.20	0.03007659	−0.02417800	−0.03365472	−0.02360602	−0.00731872	0.00618830	0.00464983	0.00547959
1.30	0.02835834	−0.02013945	−0.03539837	−0.02220405	−0.00650031	0.01371342	0.00469900	0.00487211
1.40	0.02691604	−0.01699995	−0.03654874	−0.02103215	−0.00529501	0.01538043	0.00461337	0.00397510
1.50	0.02568898	−0.01455234	−0.03724117	−0.02003865	−0.00418524	0.01127768	0.00440586	0.00314673
1.60	0.02463052	−0.01263142	−0.03758532	−0.01918423	−0.00356229	0.00354870	0.00412255	0.00268067
1.70	0.02370516	−0.01110760	−0.03766927	−0.01843921	−0.00359477	−0.00454274	0.00382652	0.00270499
1.80	0.02288616	−0.00988284	−0.03756159	−0.01778133	−0.00418850	−0.00989858	0.00357557	0.00314922
1.90	0.02215322	−0.00888675	−0.03731304	−0.01719378	−0.00504495	−0.01071085	0.00340315	0.00378885
2.00	0.02149056	−0.00807162	−0.03695903	−0.01666355	−0.00578898	−0.00701294	0.00330951	0.00434332

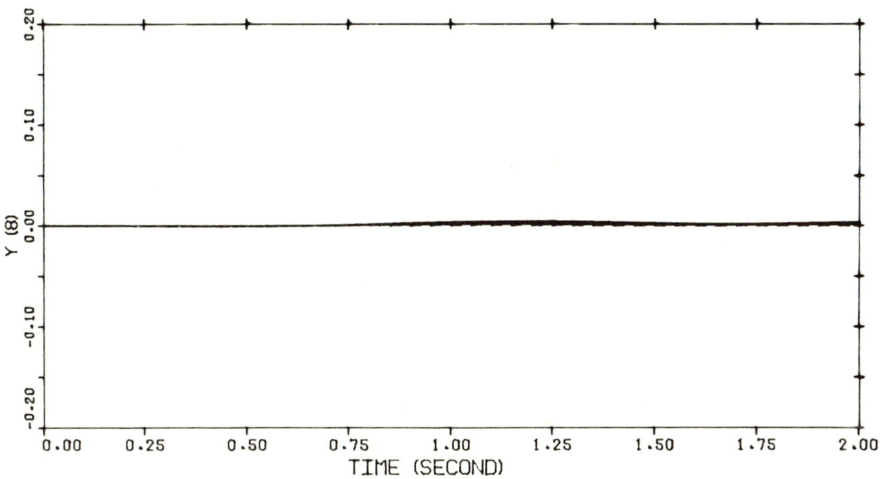

Figure 6.11. State variable versus time $Y(8)$.

Because of the restrictions on the form of the control mentioned above, additional constraints have to be considered in the implementations of the optimal solution as compared with that in Section 6.2.

The method of solution is to cast the system into the minimum-norm formulation (Ref. 6.8). The optimizing equations for the system are obtained by using the minimum-norm formulation (Ref. 6.8). Then, using the gradient descent method (Ref. 6.10) the optimal feedback parameters are obtained by solving the nonlinear two-point boundary value problem on the digital computer.

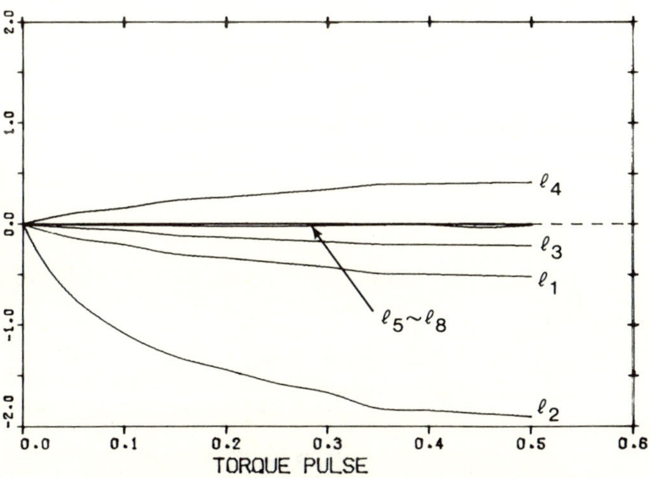

Figure 6.12. Feedback parameter versus torque pulse.

Table 6.4. Optimal Parameters, $K = 0.50$ and $x_3 = 0.10$

$e1 = -0.50380605$	$e2 = -1.92727375$	$e3 = -0.21630991$	$e4 = 0.39863914$
$f1 = -0.27903068$	$f2 = 0.57855546$	$f3 = -0.06941873$	$f4 = 0.22823286$
$g1 = -0.00002062$	$g2 = -0.0043299$	$g3 = 0.00010650$	$g4 = 0.00001558$
$h1 = -0.00046141$	$h2 = -0.00441594$	$h3 = -0.00159954$	$h4 = 0.00034891$
$a1 = -0.00011268$	$a2 = -0.00004009$	$a3 = 0.00000133$	$a4 = 0.00008987$
$b1 = 0.00005898$	$b2 = -0.0018181$	$b3 = 0.00000574$	$b4 = -0.00004859$
$c1 = -0.00000006$	$c2 = 0.00000122$	$c3 = -0.00000021$	$c4 = 0.00000004$
$d1 = 0.00000249$	$d2 = 0.00002367$	$d3 = 0.00000671$	$d4 = -0.00000188$

6.4.1. Mathematical Model

The model used in this section is a specialized form of the mathematical model used in Section 6.2.

Inclusion of the transfer functions for the field, the governor, and the turbine for each machine adds three additional differential equations to those in Section 6.2 as shown in Ref. 6.13. Here, the same procedure is used as in Ref. 6.13. The machine equations for the two machines, together with the differential equations generated by the transfer functions, are as follows:

$$\dot{\delta}_1 = \eta_1 \tag{6.323}$$

$$\dot{\eta}_1 = a_1 u_{d1} u_{q1} + b_1 u_{d1}(\omega_0 \psi_{fd1}) + (1/M_1)x_5 - (1/M_1)K_{d1}\eta_1 \tag{6.324}$$

$$\omega_0 \dot{\psi}_{fd1} = x_6 + c_1 u_{q1} - d_1(\omega_0 \psi_{fd1}) \tag{6.325}$$

$$\dot{x}_5 = (1/T_{b1})(x_7 - x_5) \tag{6.326}$$

$$\dot{x}_6 = (1/T_{e1})(C_{r1} V_2 - x_6) \tag{6.327}$$

$$\dot{x}_7 = (1/T_{g1})(G_2 G_3 V_1 - x_7) \tag{6.328}$$

$$\dot{\delta}_2 = \eta_2 \tag{6.329}$$

$$\dot{\eta}_2 = a_2 u_{d2} u_{q2} + b_2 u_{d2}(\omega_0 \psi_{fd2}) + (1/M_2)x_{12} - (1/M_2)K_{d2}\eta_2 \tag{6.330}$$

$$\omega_0 \dot{\psi}_{fd2} = x_{13} + c_2 u_{q2} - d_2 x_7 \tag{6.331}$$

$$\dot{x}_{12} = (1/T_{b2})(x_{14} - x_{12}) \tag{6.332}$$

$$\dot{x}_{13} = (1/T_{e2})(C_{r2} V_4 - x_{13}) \tag{6.333}$$

$$\dot{x}_{14} = (1/T_{g2})(G_5 G_6 V_3 - x_{14}) \tag{6.334}$$

Table 6.5. State Space Variables, $K = 0.50$ and $x_3 = 0.01$

t	$Y(1)$	$Y(2)$	$Y(3)$	$Y(4)$	$Y(5)$	$Y(6)$	$Y(7)$	$Y(8)$
0.0	0.00000000	0.00000000	0.00000000	0.00000000	0.00000000	0.00000000	0.00000000	0.00000000
0.05	0.00484641	0.18571550	0.00347477	−0.00368322	0.00000204	0.00025575	0.00002608	−0.00000154
0.10	0.01232104	0.186k6897	0.00824897	−0.00945479	0.00003065	0.00144323	0.00015817	−0.00002317
0.15	0.01941084	0.17318523	0.01201840	−0.01503468	0.00011965	0.00326097	0.00039415	−0.00009044
0.20	0.02594201	0.16051090	0.01473249	−0.02026788	0.00028450	0.00516749	0.00071127	−0.00021510
0.25	0.03196144	0.14911991	0.01650834	−0.02517158	0.00051942	0.00665944	0.00108784	−0.00039283
0.30	0.03752671	0.13896155	0.01746308	−0.02977538	0.00080020	0.00733298	0.00150553	−0.00060540
0.35	0.04269004	0.12991291	0.01770147	−0.03410793	0.00108885	0.00691613	0.00194995	−0.00082408
0.40	0.04749758	0.12855680	0.01731598	−0.03819589	0.00133922	0.00528591	0.00241071	−0.00101390
0.45	0.05199003	0.11468244	0.01638791	−0.04206368	0.00150301	0.00247048	0.00288133	−0.00113815
0.50	0.05620326	0.10829628	0.01498859	−0.04573366	0.00153564	−0.00136321	0.00335882	−0.00116291
0.60	0.05228851	−0.08752388	0.00283013	−0.04231847	0.00103713	−0.01275064	0.00417064	−0.00078586
0.70	0.04630145	−0.06956047	−0.00864566	−0.03716491	−0.00052927	−0.02706973	0.00462538	0.00039991
0.80	0.04155694	−0.05480051	−0.01723860	−0.03313922	−0.00310897	−0.03664566	0.00492144	0.00234073
0.90	0.03778865	−0.04350115	−0.02364136	−0.02997799	−0.00611188	−0.03651659	0.00520961	0.00458363
1.00	0.03477525	−0.03481295	−0.02837966	−0.02747276	−0.00871247	−0.02623029	0.00554153	0.00651187
1.10	0.03234667	−0.02811828	−0.03183898	−0.02546828	−0.01020657	−0.00943870	0.00588280	0.00761373
1.20	0.03037079	−0.02296576	−0.03430428	−0.02384695	−0.01028533	0.00792162	0.00615066	0.00767167
1.30	0.02874409	−0.01901419	−0.03599179	−0.02251851	−0.00913312	0.02005061	0.00626099	0.00682265
1.40	0.02738550	−0.01599452	−0.03707280	−0.02141377	−0.00723925	0.02341248	0.00616819	0.00548838
1.50	0.02623208	−0.01368791	−0.03768873	−0.02047825	−0.00560890	0.01786601	0.00588576	0.00421003
1.60	0.02523600	−0.01191469	−0.03795855	−0.01967297	−0.00458684	0.00646926	0.00548216	0.00344783
1.70	0.02436179	−0.01052982	−0.03798049	−0.01896796	−0.00589068	−0.00589069	0.00505538	0.00342011
1.80	0.02358387	−0.00942193	−0.03783076	−0.01834198	−0.00538577	−0.01445263	0.00469736	0.00404394
1.90	0.02288396	−0.00851327	−0.03756279	−0.01777987	−0.00666519	−0.01632620	0.00446280	0.00499592
2.00	0.02224961	−0.00775693	−0.03720910	−0.01727049	−0.00782772	−0.01137542	0.00435415	0.00585818

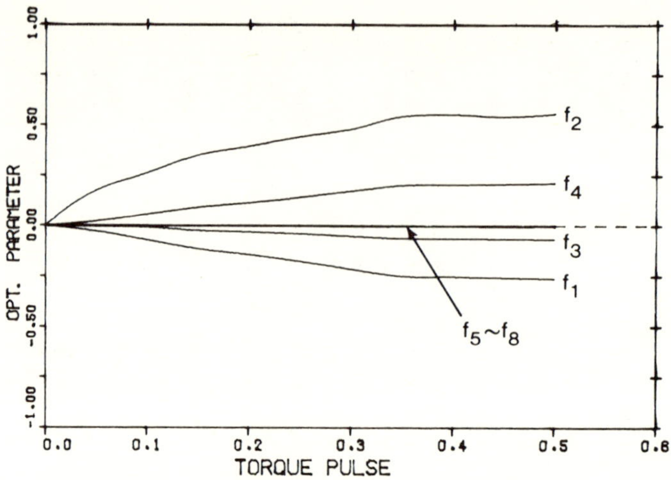

Figure 6.13. Feedback parameter versus torque pulse.

Table 6.6. Optimal Parameters, $K = 0.50$ and $x_3 = 0.01$

$e1 = -0.50109380$	$e2 = -1.92287064$	$e3 = -0.21585560$	$e4 = 0.39648038$
$f1 = -0.27771783$	$f2 = 0.59860474$	$f3 = -0.06815976$	$f4 = 0.22723585$
$g1 = -0.00016161$	$g2 = -0.00170446$	$g3 = -0.00018306$	$g4 = 0.00012225$
$h1 = -0.00058858$	$h2 = -0.00565863$	$h3 = -0.00184933$	$h4 = 0.00044513$
$a1 = -0.00009512$	$a2 = -0.00002439$	$a3 = 0.00001560$	$a4 = 0.00007564$
$b1 = 0.00009368$	$b2 = -0.00016140$	$b3 = 0.00001766$	$b4 = -0.00007663$
$c1 = 0.00000915$	$c2 = 0.00008124$	$c3 = 0.00001801$	$c4 = -0.00000692$
$d1 = 0.00000692$	$d2 = 0.00006796$	$d3 = 0.00001297$	$d4 = -0.00000524$

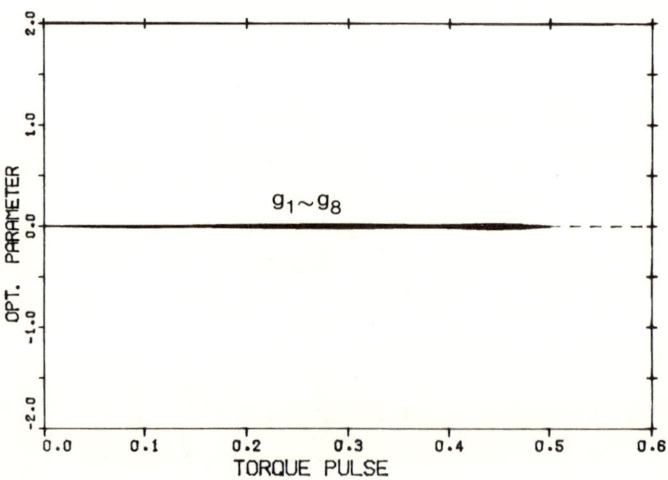

Figure 6.14. Feedback parameter versus torque pulse.

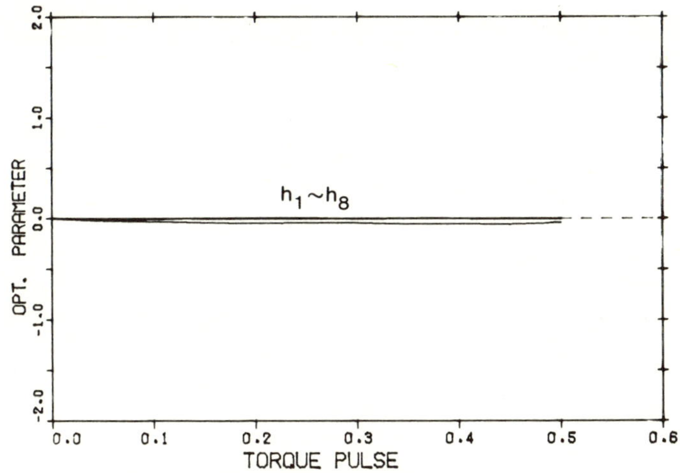

Figure 6.15. Feedback parameter versus torque pulse.

and

$$\begin{aligned} u_{d1} = {} & \{1/[A^* + B^* \cos^2(\delta_2 - \delta_1)]\} \\ & \times [-E_1^* F_1^* \sin \delta_1 - (E_1^* F_2^* + E_2^* F_2^*) \sin \delta_2 \cos(\delta_1 - \delta_2) \\ & + E_1^* F_3^* F_4^* \cos \delta_2 \sin(\delta_1 - \delta_2) + \beta_2 E_1^* F_3^* \omega_0 \psi_{fd2} \sin(\delta_1 - \delta_2) \\ & + \beta_1 E_2^* \omega_0 \psi_{fd1} \sin(\delta_1 - \delta_2) \cos(\delta_1 - \delta_2)] \end{aligned} \tag{6.335}$$

$$\begin{aligned} u_{q1} = {} & \{1/[A^* + B^* \cos^2(\delta_2 - \delta_1)]\}[-E_3^* F_1^* \cos \delta_1 \\ & + (E_2^* F_3^* F_4^* + E_3^* F_3^* F_4^* - E_2^* F_1^*) \cos \delta_2 \cos(\delta_1 - \delta_2) \\ & - E_3^* \beta_1 \omega_0 \psi_{fd1} + E_3^* F_2^* \sin \delta_2 \sin(\delta_1 - \delta_2) \\ & - E_2^* \beta_1 \omega_0 \psi_{fd1} \cos^2(\delta_1 - \delta_2) \\ & + (E_2^* + E_3^*) F_3^* \beta_2 \omega_0 \psi_{fd2} \cos(\delta_1 - \delta_2)] \end{aligned} \tag{6.336}$$

$$\begin{aligned} u_{d2} = {} & \{1/[A + B \cos^2(\delta_2 - \delta_1)]\} \\ & \times [-E_1 F_1 \sin \delta_2 - (E_1 F_2 + E_2 F_1 + E_2 F_2) \sin \delta_1 \cos(\delta_2 - \delta_1) \\ & + E_1 F_3 F_4 \cos \delta_1 \sin(\delta_2 - \delta_1) + \beta_1 E_1 F_3 \omega_0 \psi_{fd1} \sin(\delta_2 - \delta_1) \\ & + \beta_2 E_2 \omega_0 \psi_{fd2} \sin(\delta_2 - \delta_1) \cos(\delta_2 - \delta_1)] \end{aligned} \tag{6.337}$$

$$\begin{aligned} u_{q2} = {} & \{1/[A + B \cos^2(\delta_2 - \delta_1)]\} \\ & \times [-E_3 F_1 \cos \delta_2 + (E_2 F_3 F_4 + E_3 F_3 F_4 - E_2 F_1) \cos \delta_1 \cos (\delta_2 - \delta_1) \\ & - E_3 \beta_2 \omega_0 \psi_{fd2} + E_3 F_2 \sin \delta_1 \sin(\delta_2 - \delta_1) \\ & - E_2 \beta_2 \omega_0 \omega_{fd2} \cos^2(\delta_2 - \delta_1) \\ & + (E_2 + E_3) F_3 \beta_1 \omega_0 \psi_{fd1} \cos(\delta_2 - \delta_1)] \end{aligned} \tag{6.338}$$

Here, δ_1, η_1, ψ_{fd} are the torque angle, the derivative of the torque angle, and the field flux; x_5, x_6, x_7 are the mechanical power, the field voltage times the sychronous speed, and the steam power. All of the above state variables are referred to the first machine. The rest of the state variables (δ_2, η_2, ψ_{fd2}, x_{12}, x_{13}, x_{14}) are referred to the second machine. Symbols other than those mentioned above are machine parameters; T_b, T_g, T_e denote the time constants of the turbine, the governor, and the exciter; the subscripts 1 and 2 are used to represent the first and the second machine; G_2, G_3, C_{r1} are the governor, the steam valve, and the exciter gain constants of the first machine; G_5, G_6, C_{r2} are the corresponding constants of the second machine.

Use is made of the following transformations:

$$x_1 = \sin \delta_1 \tag{6.339}$$

$$x_2 = \eta_1 \tag{6.340}$$

$$x_3 = \omega_0 \psi_{fd1} \tag{6.341}$$

$$x_4 = \cos \delta_1 \tag{6.342}$$

$$x_8 = \sin \delta_2 \tag{6.343}$$

$$x_9 = \eta_2 \tag{6.344}$$

$$x_{10} = \omega_0 \psi_{fd2} \tag{6.345}$$

$$x_{11} = \cos \delta_2 \tag{6.346}$$

Then, the machine equations become

$$\dot{x}_1 = x_2 x_4 \tag{6.347}$$

$$\dot{x}_2 = a_1 u_{d1} u_{q1} + b_1 u_{d1} x_3 + (1/M_1) x_5 - (1/M_1) K_{d1} x_2 \tag{6.348}$$

$$\dot{x}_3 = x_6 + c_1 u_{q1} - d_1 x_3 \tag{6.349}$$

$$\dot{x}_4 = -x_1 x_2 \tag{6.350}$$

$$\dot{x}_8 = x_9 x_{11} \tag{6.351}$$

$$\dot{x}_9 = a_2 u_{d2} u_{q2} + b_2 u_{d2} x_{10} + (1/M_2) x_{12} - (1/M_2) K_{d2} x_9 \tag{6.352}$$

$$\dot{x}_{10} = x_{13} + c_2 u_{q2} - d_2 x_{10} \tag{6.353}$$

$$\dot{x}_{11} = -x_8 x_9 \tag{6.354}$$

Equations (6.335)–(6.338) are somewhat awkward, because they make the system nonpolynomic in view of the terms

$$1/[A^* + B^* \cos^2(\delta_2 - \delta_1)] \quad \text{and} \quad 1/[A + B \cos^2(\delta_2 - \delta_1)]$$

Noting that (see Appendix 6.A)

$$A = A^* \simeq -140.55, \qquad B = B^* \simeq 3.3$$

we have

$$1/[A^* + B^* \cos^2(\delta_2 - \delta_1)] \simeq 1/(A^* + B^*)$$
$$1/[A + B \cos^2(\delta_2 - \delta_1)] \simeq 1/(A + B)$$

Furthermore, only deviations of the state variables from steady state are considered and the two machines are equally loaded; so, we assume that

$$\sin(\delta_2 - \delta_1) \simeq 0, \qquad \cos(\delta_2 - \delta_1) \simeq 1$$

Using the relations

$$\sin \delta_1 \cos(\delta_2 - \delta_1) + \cos \delta_1 \sin(\delta_2 - \delta_1) = \sin \delta_2$$
$$\cos \delta_1 \cos(\delta_1 - \delta_2) + \sin \delta_1 \sin(\delta_1 - \delta_2) = \cos \delta_2$$

we have

$$\begin{aligned} u_{d1} &\simeq R_1^* \sin \delta_1 + R_2^* \sin \delta_2 \\ u_{q1} &\simeq S_1^* \cos \delta_1 + S_2^* \cos \delta_2 + S_3^* x_1 + S_4^* x_2 \\ u_{d2} &\simeq R_1 \sin \delta_1 + R_2 \sin \delta_2 \\ u_{q2} &\simeq S_1 \cos \delta_1 + S_2 \cos \delta_2 + S_3 x_1 + S_4 x_2 \end{aligned} \tag{6.355}$$

where

$$\begin{aligned} R_1^* &= E_1^*(F_3^* F_4^* - F_1^*)/(A^* + B^*) \\ R_2^* &= -(E_1^* F_2^* + E_2^* F_1^* + E_2^* F_2^* + E_1^* F_3^* F_4^*)/(A^* + B^*) \\ S_1^* &= -E_3^*(F_1^* + F_2^*)/(A^* + B^*) \\ S_2^* &= (E_2^* F_3^* F_4^* + E_3^* F_3^* F_4^* + E_3^* F_2^* - E_2^* F_1^*)/(A^* + B^*) \\ S_3^* &= -\beta_1(E_2^* + E_3^*)/(A^* + B^*) \\ S_4^* &= \beta_2 F_3^*(E_2^* + E_3^*)/(A^* + B^*) \\ R_1 &= -(E_1 F_2 + E_2 F_1 + E_2 F_2 + E_1 F_3 F_4)/(A + B) \\ R_2 &= E_1(F_3 F_4 - F_1)/(A + B) \\ S_1 &= (E_2 F_3 F_4 + E_3 F_3 F_4 + E_3 F_2 - E_2 F_1)/(A + B) \\ S_2 &= -E_3(F_1 + F_2)/(A + B) \\ S_3 &= \beta_1 F_3(E_2 + E_3)/(A + B) \\ S_4 &= -\beta_2(E_2 + E_3)/(A + B) \end{aligned} \tag{6.356}$$

The cost functional is given by

$$J = \int_{t_0}^{t_f} \left[\sum_{i=1}^{14} k_i(x_i - x_i^s)^2 + \sum_{i=1}^{4} l_i(V_i - V_i^s)^2 \right] dt \tag{6.357}$$

Notice that additional quadratic penalties on x_5, x_6, x_7, x_{12}, x_{13}, x_{14} have been included.

For this problem, the control is specialized to be of the form

$$V_1 - V_1^s = \sum_{i=1}^{4} e_i(x_i - x_i^s) + \sum_{i=5}^{8} e_i(x_{i+3} - x_{i+3}^s) \tag{6.358}$$

$$V_2 - V_2^s = \sum_{i=1}^{4} f_i(x_i - x_i^s) + \sum_{i=5}^{8} f_i(x_{i+3} - x_{i+3}^s) \tag{6.359}$$

$$V_3 - V_3^s = \sum_{i=1}^{4} g_i(x_i - x_i^s) + \sum_{i=5}^{8} g_i(x_{i+3} - x_{i+3}^s) \tag{6.360}$$

$$V_4 - V_4^s = \sum_{i=1}^{4} h_i(x_i - x_i^s) + \sum_{i=5}^{8} h_i(x_{i+3} - x_{i+3}^s) \tag{6.361}$$

where all the control parameters e_i's, f_i's, g_i's, h_i's are constants. This type of control is called *global control* (Ref. 6.11). The special case for which *local control* (Ref. 6.11) is obtained has the following form:

$$V_1 - V_1^s = \sum_{i=1}^{4} e_i(x_i - x_i^s) \tag{6.362}$$

$$v_2 - V_2^s = \sum_{i=1}^{4} f_i(x_i - x_i^s) \tag{6.363}$$

$$V_3 - V_3^s = \sum_{i=1}^{4} g_i(x_{i+7} - x_{i+7}^s) \tag{6.364}$$

$$V_4 - V_4^s = \sum_{i=1}^{4} h_i(x_{i+7} - x_{i+7}^s) \tag{6.365}$$

Note that the magnitude of the control parameters depends in general on the strength of the disturbance. This can be stated mathematically as follows:

$$\dot{e}_i = \dot{f}_i = \dot{g}_i = \dot{h}_i = 0, \qquad i = 1, 2, \ldots, 8 \qquad \text{for global control}$$

$$\dot{e}_i = \dot{f}_i = \dot{g}_i = \dot{h}_i = 0, \qquad i = 1, 2, \ldots, 4 \qquad \text{for local control}$$

It is assumed that the disturbance is modeled by introducing a torque pulse in the right-hand side of the torque equation of the first machine, which has magnitude K and duration τ. Hence, in this model, the values of the variables at $t = t_0$ are the steady-state values. Therefore, the torque equation of the first machine for this model is modified to

$$\dot{x}_2 = a_1 u_{d1} u_{q1} + b_1 u_{d1} x_3 + (1/M_1)\{x_5 - x_2 + K[u(t - t_0) - u(t - \tau)]\} \tag{6.366}$$

where

$$u(t) = \begin{cases} 1, & t > 0 \\ 0, & t \leq 0 \end{cases} \tag{6.367}$$

is a step function.

6.4.2. Mathematical Solution

We substitute

$$\begin{aligned} y_i &= x_i - x_i^s, & i &= 1, 2, \ldots, 14 \\ v_j &= V_j - V_j^s, & j &= 1, 2, \ldots, 4 \end{aligned} \tag{6.368}$$

where y_i and v_j are deviations from the steady state, into the cost functional (6.357) and into equations (6.347)-(6.354), (6.358)-(6.361), and (6.362)-(6.365). We obtain

$$J = \int_{t_0}^{t_f} \left[\sum_{i=1}^{14} k_i y_i^2 + \sum_{i=1}^{8} (m_i e_i^2 + n_i f_i^2 + h_i g_i^2 + q_i h_i^2) + \sum_{i=1}^{4} l_i v_i^2 \right] dt \tag{6.369}$$

Note that x_2^s and x_9^s are put equal to zero, because the torque angles do not change in the steady state. Also, quadratic penalties are imposed on the e_i's, f_i's, g_i's, h_i's. We have

$$\dot{y}_1 = x_4^s y_2 + y_2 y_4 \tag{6.370}$$

$$\begin{aligned} \dot{y}_2 = {} & a_1(u_{d1}^s u_{q1}^y + u_{q1}^s u_{d1}^y + u_{d1}^y u_{q1}^y) + b_1(x_3^s u_{d1}^y + u_{d1}^s + u_{d1}^s y_3) \\ & + (1/M_1)\{y_5 - K_{d1} y_2 + K[u(t - t_0) - u(t - \tau)]\} \end{aligned} \tag{6.371}$$

$$\dot{y}_3 = y_6 + C_1 u_{q1}^y - d_1 y_3 \tag{6.372}$$

$$\dot{y}_4 = -x_1^s y_2 - y_1 y_2 \tag{6.373}$$

$$\dot{y}_5 = (1/T_{b1})(y_1 - y_5) \tag{6.374}$$

$$\dot{y}_6 = (1/T_{e1})(C_{r1} v_2 - y_6) \tag{6.375}$$

$$\dot{y}_7 = (1/T_{g1})(G_2 G_3 v_1 - y_7) \tag{6.376}$$

$$\dot{y}_8 = x_{11}^s v_9 + y_9 y_{11} \tag{6.377}$$

$$\dot{y}_9 = a_2(u^s_{d2}u^y_{q2} + u^s_{q2}u^y_{d2} + u^y_{d2}u^y_{q2}) + b_2(x^s_{10}u^y_{d2} + u^s_{d2}y_{10} + u^y_{d2}y_{10}) + (1/M_2)(y_{12} - K_{d2}y_9) \tag{6.378}$$

$$\dot{y}_{10} = y_{13} + C_2u^y_{q2} - d_2y_{10} \tag{6.379}$$

$$\dot{y}_{11} = -x^s_8y_9 - y_8y_9 \tag{6.380}$$

$$\dot{y}_{12} = (1/T_{b2})(y_{14} - y_{12}) \tag{6.381}$$

$$\dot{y}_{13} = (1/T_{e2})(C_{r2}v_4 - y_{13}) \tag{6.382}$$

$$\dot{y}_{14} = (1/T_{g2})(G_5G_6v_3 - y_{14}) \tag{6.383}$$

For global control, we have

$$v_1 = \sum_{i=1}^{4} e_iy_i + \sum_{i=5}^{8} e_iy_{i+3} \tag{6.384}$$

$$v_2 = \sum_{i=1}^{4} f_iy_i + \sum_{i=5}^{8} f_iy_{i+3} \tag{6.385}$$

$$v_3 = \sum_{i=1}^{4} g_iy_i + \sum_{i=5}^{8} g_iy_{i+3} \tag{6.386}$$

$$v_4 = \sum_{i=1}^{4} h_iy_i + \sum_{i=5}^{8} h_iy_{i+3} \tag{3.387}$$

$$\dot{e}_i = \dot{f}_i = \dot{g}_i = \dot{h}_i = 0, \qquad i = 1, 2, \ldots, 8 \tag{3.388}$$

For local control, we have

$$v_1 = \sum_{i=1}^{4} e_iy_i \tag{6.389}$$

$$v_2 = \sum_{i=1}^{4} f_iy_i \tag{6.390}$$

$$v_3 = \sum_{i=1}^{4} g_iy_{i+7} \tag{6.391}$$

$$v_4 = \sum_{i=1}^{4} h_iy_{i+7} \tag{6.392}$$

$$\dot{e}_i = \dot{f}_i = \dot{g}_i = \dot{h}_i = 0, \qquad i = 1, 2, \ldots, 4 \tag{6.393}$$

It should be noted that from equations (6.368) and (6.370)–(6.383) it follows that, at steady state,

$$x^s_2 = 0$$

$$x^s_9 = 0$$

$$a_1 u_{d1}^s u_{q1}^s + b_1 u_{d1}^s x_3^s + x_5^s = 0$$

$$a_2 u_{d2}^s u_{q2}^s + b_2 u_{d2}^s x_{10}^s + x_{12}^s = 0$$

$$x_6^s + c_1 u_{q1}^s - d_1 x_3^s = 0$$

$$x_{13}^s + c_2 u_{q2}^2 - d_2 x_{10}^s = 0$$

$$x_5^s = x_7^s$$

$$C_{r1} V_2^s - x_6^s = 0$$

$$G_2 G_3 V_1^s - x_7^s = 0$$

$$x_{12}^s = x_{14}^s$$

$$C_{r2} V_4^s - x_{13}^s = 0$$

$$G_5 G_6 V_3^s - x_{14}^s = 0$$

The initial values are

$$y_i(t_0) = 0, \qquad i, 1, 2, \ldots, 14 \tag{6.394}$$

Then, the augmented cost functional is

$$J_1 = \int_{t_0}^{t_f} F_G \, dt \qquad \text{for global control} \tag{6.395}$$

$$J_1 = \int_{t_0}^{t_f} F_L \, dt \qquad \text{for local control} \tag{6.396}$$

in the above equations,

$$\begin{aligned}
F_G = {} & \sum_{i=1}^{14} k_i y_i^2 + \sum_{i=1}^{8} (m_i e_i^2 + n_i f_i^2 + p_i g_i^2 + q_i h_i^2) + \sum_{i=1}^{4} l_i v_i^2 \\
& + \mu_1\left[v_1 - \sum_{i=1}^{4} e_i y_i - \sum_{i=5}^{8} e_i y_{i+3}\right] + \mu_2\left[v_2 - \sum_{i=1}^{4} f_i y_i - \sum_{i=5}^{8} f_i y_{i+3}\right] \\
& + \mu_3\left[v_3 - \sum_{i=1}^{4} g_i y_i - \sum_{i=5}^{8} g_i y_{i+3}\right] + \mu_4\left[v_4 - \sum_{i=1}^{4} h_i y_i - \sum_{i=5}^{8} h_i y_{i+3}\right] \\
& + \sum_{i=1}^{8} \nu_i \dot{e}_i + \sum_{i=1}^{8} \nu_{i+8} \dot{f}_i + \sum_{i=1}^{8} \nu_{i+16} \dot{g}_i + \sum_{i=1}^{8} \nu_{i+24} \dot{h}_i \\
& + \lambda_1[\dot{y}_1 - (x_4^s y_2 + y_2 y_4)] + \lambda_2\{\dot{y}_2 - a_1(u_{d1}^s u_{q1}^y + u_{q1}^s u_{d1}^y + u_{d1}^y u_{q1}^y) \\
& - b_1(x_3^s u_{d1}^y + u_{d1}^s y_3 + u_{d1}^y y_3) \\
& - (1/M_1)[y_5 - K_{d1} y_2 + K(u(t - t_0) - u(t - \tau))]\} \\
& + \lambda_3(\dot{y}_3 - y_6 - G u_{q1}^y + d_1 y_3) + \lambda_4(\dot{y}_4 + x_1^s y_2 + y_1 y_2)
\end{aligned}$$

$$\begin{aligned}
&+ \lambda_5[\dot{y}_5 - (1/T_{b1})(y_7 - y_5)] + \lambda_6[\dot{y}_6 - (1/T_{e1})(C_{r1}v_2 - y_6)] \\
&+ \lambda_7[\dot{y}_7 - (1/T_{g1})(G_2G_3v_1 - y_7)] + \lambda_8(\dot{y}_8 - x_{11}^s y_9 - y_9 y_{11}) \\
&+ \lambda_9[(\dot{y}_9 - a_2(u_{d2}^s u_{q2}^y + u_{q1}^s u_{d2}^y + u_{d2}^y u_{q2}^y)) \\
&- b_2(x_{10}^s u_{d2}^y + u_{d2}^s y_{10} + u_{d2}^y y_{10}) - (1/M_2)(y_{12} - K_{d2}y_9)] \\
&+ \lambda_{10}(\dot{y}_{10} - y_{13} - c_2 u_{q2}^y - d_2 y_{10}) + \lambda_{11}(\dot{y}_{11} + x_8^s y_9 + y_8 y_9) \\
&+ \lambda_{12}[\dot{y}_{12} - (1/T_{b2})(y_{14} - y_{12})] \\
&+ \lambda_{13}[\dot{y}_{13} - (1/T_{e2})(C_{r2}v_4 - y_{13})] \\
&+ \lambda_{14}[\dot{y}_{14} - (1/T_{g2})(G_5G_6v_3 - y_{14})]
\end{aligned} \tag{6.397}$$

and

$$\begin{aligned}
F_L &= \sum_{i=1}^{14} k_i y_i^2 + \sum_{i=1}^{4} (m_i e_i^2 + n_i f_i^2 + p_i g_i^2 + q_i h_i^2) + \sum_{i=1}^{4} l_i v_i^2 \\
&+ \mu_1\left[v_1 - \sum_{i=1}^{4} e_i y_i\right] + \mu_2\left[v_2 - \sum_{i=1}^{4} f_i y_i\right] \\
&+ \mu_3\left[v_3 - \sum_{i=1}^{4} g_i y_{i+7}\right] + \mu_4\left[v_4 - \sum_{i=1}^{4} h_i y_{i+7}\right] \\
&+ \sum_{i=1}^{4} \nu_i \dot{e}_i + \sum_{i=1}^{4} \nu_{i+4}\dot{f}_i + \sum_{i=1}^{4} \nu_{i+8}\dot{g}_i + \sum_{i=1}^{4} \nu_{i+12}\dot{h}_i \\
&+ \lambda_1[\dot{y}_1 - (x_4^s y_2 + y_2 y_4)] \\
&+ \lambda_2\{\dot{y}_2 - a_1(u_{d1}^s u_{q1}^y + u_{q1}^s u_{d1}^y + u_{d1}^y u_{q1}^y) - b_1(x_3^s u_{d1}^y + u_{d1}^s y_3 + u_{d1}^y y_3) \\
&- (1/M_1)[y_5 - K_{d1}y_2 + K(u(t - t_0) - u(t - \tau))]\} + \cdots
\end{aligned} \tag{6.398}$$

In (6.398), the terms with multipliers $\lambda_3, \lambda_4, \ldots, \lambda_{14}$ have not been written, since they are identical to the corresponding terms in (6.397).

Integrating by parts, collecting terms that are quadratic and linear in the u_i's, e_i's, f_i's, g_i's, h_i's, and dropping terms that do not depend on these, the augmented cost functional reduces to

$$J_1 = \int_{t_0}^{t_f} (Y_G^T R_G Y_G + L_G^T Y_G)\, dt + \text{boundary terms, for global control} \tag{6.399}$$

Here, T denotes transpose,

$$Y = (y_1 \sim y_{14}, e_1 \sim e_8, f_1 \sim f_8, g_1 \sim g_8, h_1 \sim h_8, v_1 \sim v_4)$$

is a vector of order 50,

$$L = (L_1 \sim L_{14}, L_{15} \sim L_{22}, L_{23} \sim L_{30}, L_{31} \sim L_{38}, L_{39} \sim L_{46}, L_{47} \sim L_{50})$$

is a vector of order 50, and R is a matrix of order 50×50, which has the following partitioned form:

$$R = \begin{bmatrix} R_{11} & R_{12} & R_{13} & R_{14} & R_{15} & 0_{14\times4} \\ R_{12}^T & M_1 & 0_{8\times8} & 0_{8\times8} & 0_{8\times8} & 0_{8\times4} \\ R_{13}^T & 0_{8\times8} & N_1 & 0_{8\times8} & 0_{8\times8} & 0_{8\times4} \\ R_{14}^T & 0_{8\times8} & 0_{8\times8} & P_1 & 0_{8\times8} & 0_{8\times4} \\ R_{15}^T & 0_{8\times8} & 0_{8\times8} & 0_{8\times8} & Q_1 & Q_{8\times4} \\ 0_{4\times14} & 0_{4\times8} & 0_{4\times8} & 0_{4\times8} & 0_{4\times8} & L_1 \end{bmatrix} \tag{6.400}$$

The components of the vector L are given by

$$L_1 = -\dot{\lambda}_1 - a_1R_1^*u_{q1}^s\lambda_2 - b_1R_1^*x_3^s\lambda_2 - (a_2u_{q2}^s + b_2x_7^s)R_1\lambda_6$$

$$L_2 = -\dot{\lambda}_2 + K_{d1}^*\lambda_2 + x_1^s\lambda_4 - x_4^s\lambda_1$$

$$L_3 = -\dot{\lambda}_3 - a_1S_3^*u_{d1}^s\lambda_2 - c_1S_3^*\lambda_3 + d_1\lambda_3 - a_2S_3u_{d2}^s\lambda_6 - c_2S_3\lambda_7$$
$$- b_1u_{d1}^s\lambda_1$$

$$L_4 = -\dot{\lambda}_4 - a_1S_1^*u_{d1}^s\lambda_2 - c_1S_1^*\lambda_3 - a_2S_1u_{d2}^s\lambda_6 - c_2S_1\lambda_7$$

$$L_5 = -\dot{\lambda}_5 + (1/T_{b1})\lambda_5 - (1/M_1)\lambda_2$$

$$L_6 = -\dot{\lambda}_6 + (1/T_{e1})\lambda_6 - \lambda_3$$

$$L_7 = -\dot{\lambda}_7 + (1/T_{g1})\lambda_7 - (1/T_{b1})\lambda_5$$

$$L_8 = -\dot{\lambda}_5 - (a_1u_{q1}^s + b_1x_3^s)R_2^*\lambda_2 - (a_2u_{q2}^s + b_2x_7^s)R_2\lambda_6$$

$$L_9 = -\dot{\lambda}_6 - x_8^s\lambda_5 + K_{d2}\lambda_6 + x_5^s\lambda_8$$

$$L_{10} = -\dot{\lambda}_7 - a_1S_4^*u_{d1}^s\lambda_2 - c_1S_4^*\lambda_3 - (a_2S_4 + b_2)u_{d2}^s\lambda_6 + (d_2 - c_2S_4)\lambda_7$$

$$L_{11} = -\dot{\lambda}_8 - a_1S_2^*u_{d1}^s\lambda_2 - c_1S_3^*\lambda_3 - a_2S_2u_{d2}^s\lambda_6 - c_2S_2\lambda_7$$

$$L_{12} = -\dot{\lambda}_{12} + (1/T_{b2})\lambda_{12} - (1/M_2)\lambda_9$$

$$L_{13} = -\dot{\lambda}_{13} + (1/T_{e2})\lambda_{13} - \lambda_{10}$$

$$L_{14} = -\dot{\lambda}_{14} + (1/T_{g2})\lambda_{14} - (1/T_{b2})\lambda_{12}$$

$$L_{K=14} = -\dot{\nu}_K, \qquad K = 1, 2, \ldots, 32$$

$$L_{47} = \mu_1 - (1/T_{g1})G_2G_3\lambda_7$$

$$L_{48} = \mu_2 - (1/T_{e1})C_{r1}\lambda_6$$

$$L_{49} = \mu_3 - (1/T_{g2})G_5G_6\lambda_{14}$$

$$L_{50} = \mu_4 - (1/T_{e2})C_{r2}\lambda_{13}$$

The symmetric matrix R_{11} is of order 14×14, and its components are given by

$$\nu_{ii} = k_i, \qquad \text{for } i = 1, 2, \ldots, 14$$

For the nondiagonal terms, we list only the elements that are not zero:

$$\gamma_{12} = \tfrac{1}{2}\lambda_4$$

$$\gamma_{13} = -\tfrac{1}{2}(a_1R_1^*S_3^*\lambda_2 + a_2R_1S_3\lambda_9 + b_1R_1^*\lambda_2)$$

$$\gamma_{14} = -\tfrac{1}{2}(a_1R_1^*S_1^*\lambda_2 + a_2R_1S_1\lambda_9)$$

$$\gamma_{10} = -\tfrac{1}{2}(a_1R_1^*S_4^*\lambda_2 + a_2R_1S_4\lambda_9 + b_2R_1\lambda_9)$$

$$\gamma_{1,11} = -\tfrac{1}{2}(a_1R_1^*S_2^*\lambda_2 + a_2R_1S_2\lambda_9)$$

$$\gamma_{24} = -\tfrac{1}{2}\lambda_1$$

$$\gamma_{38} = -\tfrac{1}{2}(a_1S_3^*R_2^*\lambda_2 + b_1R_2^*\lambda_2 + a_2R_2S_3\lambda_9)$$

$$\gamma_{48} = -\tfrac{1}{2}(a_1R_2^*S_1^*\lambda_2 + a_2R_2S_1\lambda_9)$$

$$\gamma_{89} = \tfrac{1}{2}\lambda_{11}$$

$$\gamma_{8,10} = -\tfrac{1}{2}(a_1R_2^*S_4^*\lambda_2 + a_2S_4R_2\lambda_9 + b_2R_2\lambda_9)$$

$$\gamma_{8,11} = -\tfrac{1}{2}(a_1R_2^*S_2^*\lambda_2 + a_2R_2S_2\lambda_9)$$

$$\gamma_{9,11} = -\tfrac{1}{2}\lambda_8$$

and

$$R_{1K} = -\tfrac{1}{2}\mu_{K-1}\begin{bmatrix} E_4 & 0_{4\times 4} \\ 0_{3\times 4} & 0_{3\times 4} \\ 0_{4\times 4} & E_4 \\ 0_{3\times 4} & 0_{3\times 4} \end{bmatrix}$$

$$M_1 = \text{diag}[m_i]$$

$$N_1 = \text{diag}[n_i]$$

$$P_1 = \text{diag}[p_i]$$

$$Q_1 = \text{diag}[q_i]$$

$$L_1 = \text{diag}[l_i]$$

The matrix E_4 is a unit matrix of order 4×4. The matrices $0_{3\times4}$, $0_{4\times4}$, $0_{14\times4}$, $0_{8\times4}$, $0_{8\times8}$, $0_{4\times14}$ are null matrices of order 3×4, 4×4, 14×4, 8×4, 4×8, 8×8, 4×14, respectively.

It will be noticed that J_1 in equation (6.399) is composed of an integral part and a boundary term which are independent of each other. To minimize J_1 in equation (6.399), one minimizes the boundary and the integral parts separately. If one defines the following

$$V_G = R_G^{-1} L_G \tag{6.401}$$

then, the integral part can be written in the following form similar to completing the squares as

$$\begin{aligned} J_1 &= \text{boundary terms for global control} \\ &\quad + \int_{t_p}^{t_f} [(Y_G + \tfrac{1}{2}V_G)^T R_G (Y_G + \tfrac{1}{2}V_G) - \tfrac{1}{4}V_G^T R_G V_G]\, dt \end{aligned} \tag{6.402}$$

To minimize the integral part, one can write this part as

$$\min_{Y_G} J_{\text{in}} = \min_{Y_G} \int_{t_0}^{t_f} [(Y_G + \tfrac{1}{2}V_G)^T R_G (Y_G + \tfrac{1}{2}V_G)]\, dt \tag{6.403}$$

because V_G is constant independent of Y_G. Equation (6.403) defines a norm. Hence, equation (6.403) can be written as

$$\min_{Y_G} J_{\text{in}} = \min_{Y_G} [\| Y_G + \tfrac{1}{2}V_G \|]_{R_G} \tag{6.404}$$

6.4.3. The Optimal Solution

There is only one solution for the problem formulated in equation (6.404); this is given by

$$[Y_G + \tfrac{1}{2}V_G] = [0] \tag{6.405}$$

Substituting from equation (6.401) into equation (6.405), one obtains

$$L_G = -2R_G Y_G \tag{6.406}$$

and the boundary term is minimized with respect to

$$y_i(t_f),\ e_i(t_0),\ e_i(t_f), f_i(t_0), f_i(t_f),\ g_i(t_0),\ g_i(t_f),\ h_i(t_0),\ h_i(t_f)$$

which yields

$$\lambda_i(t_f) = 0, \qquad i = 1, 2, \ldots, 14 \tag{6.407}$$

$$\nu_j(t_0) = \nu_j(t_f) = 0, \qquad j = 1, 2, \ldots, 32 \tag{6.408}$$

The optimizing equations (6.406) are arranged and given specifically as follows:

$$\begin{aligned}\dot{\lambda}_1 = 2k_1y_1 &- \lambda_2[a_1R_1^*(u_{q1}^s + u_{q1}^y) + b_1R_1^*(x_3^s + y_3)] + \lambda_4y_2\\ &- \lambda_9[a_2R_1(u_{q2}^s + u_{q2}^y) + b_2R_1(x_{10}^s + y_{10})]\\ &- (\mu_1e_1 + u_2f_1 + \mu_3g_1 + \mu_4h_1)\end{aligned} \tag{6.409}$$

$$\begin{aligned}\dot{\lambda}_2 = 2k_2y_2 &+ \lambda_4(x_1^s + y_1) - \lambda_1(x_4^s + y_4) + \lambda_2(1/M_1)K_{d1}\\ &- (\mu_1e_2 + \mu_2f_2 + \mu_3g_2 + \mu_4h_2)\end{aligned} \tag{6.410}$$

$$\begin{aligned}\dot{\lambda}_3 = 2k_3y_3 &- \lambda_2[a_1S_3^*(u_{d1}^s + u_{d1}^y) + b_1(u_{d1}^s + u_{d1}^y)] - \lambda_3(c_1S_3^* - d_1)\\ &- \lambda_9a_2S_3(u_{d2}^s + u_{d2}^y) - \lambda_{10}c_2S_3 - (\mu_1e_3 + \mu_2f_3 + \mu_3g_3 + \mu_4h_3)\end{aligned} \tag{6.411}$$

$$\begin{aligned}\dot{\lambda}_4 = 2k_4y_4 &- \lambda_1y_2 - \lambda_2a_1S_1^*(u_{d1}^s + u_{d1}^y) - \lambda_3c_1S_1^* - \lambda_9a_2S_1(u_{d2}^s + u_{d2}^y)\\ &- \lambda_{10}c_2S_1 - (\mu_1e_4 + \mu_2f_4 + \mu_3g_4 + \mu_4h_4)\end{aligned} \tag{6.412}$$

$$\dot{\lambda}_5 = 2k_5y_5 - \lambda_2(1/M_1) + \lambda_5(1/T_{b1}) \tag{6.413}$$

$$\dot{\lambda}_6 = 2k_6y_6 - \lambda_3 + \lambda_6(1/T_{e1}) \tag{6.414}$$

$$\dot{\lambda}_7 = 2k_7y_7 - \lambda_5(1/T_{b1}) + \lambda_7(1/T_{g1}) \tag{6.415}$$

$$\begin{aligned}\dot{\lambda}_8 = 2k_8y_8 &- \lambda_2[a_1R_2^*(u_{q1}^s + u_{q1}^y) + b_1R_2^*(x_3^s + y_3)]\\ &- \lambda_9[a_2R_2(u_{q2}^s + u_{q2}^y) + b_2R_2(x_{10}^s + y_{10})] + \lambda_{11}y_9\\ &- (\mu_1e_5 + \mu_2f_5 + \mu_3g_5 + \mu_4h_5)\end{aligned} \tag{6.416}$$

$$\begin{aligned}\dot{\lambda}_9 = 2k_9y_9 &- \lambda_8(x_{11}^s + y_{11}) + \lambda_9(1/M_2)K_{d2} + \lambda_{11}(x_8^s + y_8)\\ &- (\mu_1e_6 + \mu_2f_6 + \mu_3g_6 + \mu_4h_6)\end{aligned} \tag{6.417}$$

$$\dot{\lambda}_{10} = 2k_{10}y_{10} - \lambda_2 a_1 S_4^*(u_{d1}^s) - \lambda_3 c_1 S_4^* - \lambda_9[a_2 S_4(u_{d2}^s + u_{d2}^y) + b_2(u_{d2}^s + u_{d2}^y)] + \lambda_{10}(d_2 - c_2 S_4) - (\mu_1 e_7 + \mu_2 f_7 + \mu_3 g_7 + \mu_4 h_7) \quad (6.418)$$

$$\dot{\lambda}_{11} = 2k_{11}y_{11} - \lambda_2 a_1 S_2^*(u_{d1}^s + u_{d1}^y) - \lambda_3 c_1 S_2^* - \lambda_8 y_9 - \lambda_9 a_2 S_2(u_{d2}^s + u_{d2}^y) - \lambda_{10} c_2 S_2 - (\mu_1 e_8 + \mu_2 f_8 + \mu_3 g_8 + \mu_4 h_8) \quad (6.419)$$

$$\dot{\lambda}_{12} = 2k_{12}y_{12} - \lambda_9(1/M_2) + \lambda_{12}(1/T_{b2}) \quad (6.420)$$

$$\dot{\lambda}_{13} = 2k_{13}y_{13} - \lambda_{10} + \lambda_{13}(1/T_{e2}) \quad (6.421)$$

$$\dot{\lambda}_{14} = 2k_{14}y_{14} - \lambda_{12}(1/T_{b2}) + \lambda_{14}(1/T_{g2}) \quad (6.422)$$

and

$$\dot{\nu}_1 = -\mu y_1 + 2m_1 e_1 \quad (6.423)$$

$$\dot{\nu}_2 = -\mu_1 y_2 + 2m_2 e_2 \quad (6.424)$$

$$\dot{\nu}_3 = -\mu_1 y_3 + 2m_3 e_3 \quad (6.425)$$

$$\dot{\nu}_4 = -\mu_1 y_4 + 2m_4 e_4 \quad (6.426)$$

$$\dot{\nu}_5 = -\mu_1 y_8 + 2m_5 e_5 \quad (6.427)$$

$$\dot{\nu}_6 = -\mu_1 y_9 + 2m_6 e_6 \quad (6.428)$$

$$\dot{\nu}_7 = -\mu_1 y_{10} + 2m_7 e_7 \quad (6.429)$$

$$\dot{\nu}_8 = -\mu_1 y_{11} + 2m_8 e_8 \quad (6.430)$$

$$\dot{\nu}_9 = -\mu_2 y_1 + 2n_1 f_1 \quad (6.431)$$

$$\dot{\nu}_{10} = -\mu_2 y_2 + 2n_2 f_2 \quad (6.432)$$

$$\dot{\nu}_{11} = -\mu_2 y_3 + 2n_3 f_3 \quad (6.433)$$

$$\dot{\nu}_{12} = -\mu_2 y_4 + 2n_4 f_4 \quad (6.434)$$

$$\dot{\nu}_{13} = -\mu_2 y_8 + 2n_5 f_5 \quad (6.435)$$

$$\dot{\nu}_{14} = -\mu_2 y_9 + 2n_6 f_6 \quad (6.436)$$

$$\dot{\nu}_{15} = -\mu_2 y_{10} + 2n_7 f_7 \quad (6.437)$$

$$\dot{\nu}_{16} = -\mu_2 y_{11} + 2n_8 f_8 \quad (6.438)$$

$$\dot{\nu}_{17} = -\mu_3 y_1 + 2p_1 q_1 \quad (6.439)$$

$$\dot{\nu}_{18} = -\mu_3 y_2 + 2p_2 g_2 \quad (6.440)$$

$$\dot{\nu}_{19} = -\mu_3 y_3 + 2p_3 g_3 \quad (6.441)$$

$$\dot{\nu}_{20} = -\mu_3 y_4 + 2p_4 g_4 \quad (6.442)$$

$$\dot{\nu}_{21} = -\mu_3 y_8 + 2p_5 g_5 \quad (6.443)$$

$$\dot{\nu}_{22} = -\mu_3 y_9 + 2p_6 g_6 \quad (6.444)$$

$$\dot{\nu}_{23} = -\mu_3 y_{10} + 2p_7 g_7 \tag{6.445}$$

$$\dot{\nu}_{24} = -\mu_3 y_{11} + 2p_8 g_8 \tag{6.446}$$

$$\dot{\nu}_{25} = -\mu_4 y_1 + 2q_1 h_1 \tag{6.447}$$

$$\dot{\nu}_{26} = -\mu_4 y_2 + 2q_2 h_2 \tag{6.448}$$

$$\dot{\nu}_{27} = -\mu_4 y_3 + 2q_3 h_3 \tag{6.449}$$

$$\dot{\nu}_{28} = -\mu_4 y_4 + 2q_4 h_4 \tag{6.450}$$

$$\dot{\nu}_{29} = -\mu_4 y_8 + 2q_5 h_5 \tag{6.451}$$

$$\dot{\nu}_{30} = -\mu_4 y_9 + 2q_6 h_6 \tag{6.452}$$

$$\dot{\nu}_{31} = -\mu_4 y_{10} + 2q_7 h_7 \tag{6.453}$$

$$\dot{\nu}_{32} = -\mu_4 y_{11} + 2q_8 h_8$$

and

$$\mu_1 = \lambda_7(1/T_{g1})G_2G_3 - 2l_1v_1 \tag{6.454}$$

$$\mu_2 = \lambda_6(1/T_{e1})C_{r1} - 2l_2v_2 \tag{6.455}$$

$$\mu_3 = \lambda_{14}(1/T_{g2})G_5G_6 - 2l_3v_3 \tag{6.456}$$

$$\mu_4 = \lambda_{13}(1/T_{e2})c_{r2} - 2l_4v_4 \tag{6.457}$$

For the case of local control, the same procedure is used to yield the optimizing equations and boundary conditions. One thus obtains the equations

$$\dot{\lambda}_1 = 2k_1y_1 - \lambda_2[a_1R_1^*(u_{q1}^s + u_{q1}^y) + b_1R_1^*(x_3^s + y_3)] + \lambda_4 y_2 - \lambda_9[a_2R_1(u_{q2}^s + u_{q2}^y) + b_2R_1(x_{10}^s + y_{10})] - (e_1\mu_1 + f_1\mu_2) \tag{6.459}$$

$$\dot{\lambda}_2 = 2k_2y_2 + \lambda_4(x_1^s + y_1) - \lambda_1(x_4^s + y_4) + \lambda_2(1/M_1)K_{d1} - (\mu_1e_2 + \mu_2f_2) \tag{6.460}$$

$$\dot{\lambda}_3 = 2k_3y_3 - \lambda_2[a_1S_3^*(u_{d1}^s + u_{d1}^y) + b_1(u_{d1}^s + u_{d1}^y)] - \lambda_3(c_1S_3^* - d_1) - \lambda_9a_2S_3(u_{d2}^s + u_{d2}^y) - \lambda_{10}c_2S_3 - (\mu_1e_3 + \mu_2f_3) \tag{6.461}$$

$$\dot{\lambda}_4 = 2k_4y_4 - \lambda_1y_2 - \lambda_2a_1S_1^*(u_{d1}^s + u_{d1}^y) - \lambda_9a_2S_1(u_{d2}^s + u_{d2}^y) - \lambda_3c_1S_1^* - \lambda_{10}c_2S_1 - (\mu_1e_4 + \mu_2f_4) \tag{6.462}$$

$$\dot{\lambda}_5 = 2k_5y_5 - \lambda_2(1/M_1) + \lambda_5(1/T_{b1}) \tag{6.463}$$

$$\dot{\lambda}_6 = 2k_6y_6 - \lambda_3 + \lambda_6(1/T_{e1}) \tag{6.464}$$

$$\dot{\lambda}_7 = 2k_7y_7 - \lambda_7 - \lambda_5(1/T_{b1}) + \lambda_7(1/T_{g1}) \tag{6.465}$$

$$\dot{\lambda}_8 = 2k_8y_8 - \lambda_2[a_1R_2^*(u_{q1}^s + u_{q1}^y) + b_1R_2^*(x_3^s + y_3)] - \lambda_9[a_2R_2(u_{q2}^s + u_{q2}^y) + b_2R_2(x_{10}^s + y_{10})] + \lambda_{11}y_9 - (\mu_3g_1 + \mu_4h_1) \tag{6.466}$$

$$\dot{\lambda}_9 = 2k_9y_9 - \lambda_8(x_{11}^s + y_{11}) + \lambda_9(1/M_2)K_{d2} + \lambda_{11}(x_8^s + y_8) - (\mu_3 g_2 + \mu_4 h_2) \tag{6.467}$$

$$\dot{\lambda}_{10} = 2k_{10}y_{10} - \lambda_2 a_1 S_4^*(u_{d1}^s + u_{d1}^y) - \lambda_3 c_1 S_4^* - \lambda_9[a_2 S_4(u_{d2}^s + u_{d2}^y) + b_2(u_{d2}^s + u_{d2}^y)] + \lambda_{10}(d_2 - c_2 S_4) - (\mu_3 g_3 + \mu_4 h_3) \tag{6.468}$$

$$\dot{\lambda}_{11} = 2k_{11}y_{11} - \lambda_2 a_1 S_2^*(u_{d1}^s + u_{d1}^y) - \lambda_3 c_1 S_2^* - \lambda_8 y_9 - \lambda_9 a_2 S_2(u_{d2}^s + u_{d2}^y) - \lambda_{10} c_2 S_2 - (\mu_3 g_4 + \mu_4 h_4) \tag{6.469}$$

$$\lambda_{12} = 2k_{12}y_{12} - \lambda_9(1/M_2) + \lambda_{12}(1/T_{b2}) \tag{6.470}$$

$$\dot{\lambda}_{13} = 2k_{13}y_{13} - \lambda_{10} + \lambda_{13}(1/T_{e2}) \tag{6.471}$$

$$\dot{\lambda}_{14} = 2k_{14}y_{14} - \lambda_{12}(1/T_{b2}) + \lambda_{14}(1/T_{g2}) \tag{6.472}$$

and

$$\dot{\nu}_1 = -\mu_1 y_1 + 2m_1 e_1 \tag{6.473}$$

$$\dot{\nu}_2 = -\mu_1 y_2 + 2m_e e_2 \tag{6.474}$$

$$\dot{\nu}_3 = -\mu_1 y_3 + 2m_3 e_3 \tag{6.475}$$

$$\dot{\nu}_4 = -\mu_1 y_4 + 2m_4 e_4 \tag{6.476}$$

$$\dot{\nu}_5 = -\mu_2 y_1 + 2n_1 f_1 \tag{6.477}$$

$$\dot{\nu}_6 = -\mu_2 y_2 + 2n_2 f_2 \tag{6.478}$$

$$\dot{\nu}_7 = -\mu_2 y_3 + 2n_3 f_3 \tag{6.479}$$

$$\dot{\nu}_8 = -\mu_2 y_4 + 2n_4 f_4 \tag{6.480}$$

$$\dot{\nu}_9 = -\mu_3 y_8 + 2p_1 g_1 \tag{6.481}$$

$$\dot{\nu}_{10} = -\mu_3 y_9 + 2p_2 g_2 \tag{6.482}$$

$$\dot{\nu}_{11} = -\mu_3 y_{10} + 2p_3 g_3 \tag{6.483}$$

$$\dot{\nu}_{12} = -\mu_3 y_{11} + 2p_4 g_4 \tag{6.484}$$

$$\dot{\nu}_{13} = -\mu_4 y_8 + 2q_1 h_1 \tag{6.485}$$

$$\dot{\nu}_{14} = -\mu_4 y_9 + 2q_2 h_2 \tag{6.486}$$

$$\dot{\nu}_{15} = -\mu_4 y_{10} + 2q_3 h_3 \tag{6.487}$$

$$\dot{\nu}_{16} = -\mu_4 y_{11} + 2q_4 h_4 \tag{6.488}$$

and

$$\mu_1 = \lambda_7(1/T_{g1})G_2 G_3 - 2l_1 v_1 \tag{6.489}$$

$$\mu_2 = \lambda_6(1/T_{e1})C_{r1} - 2l_2v_2 \tag{6.490}$$

$$\mu_3 = \lambda_{14}(1/T_{g1})G_5G_6 - 2l_3v_3 \tag{6.491}$$

$$\mu_4 = \lambda_{13}(1/T_{e2})C_{r2} - 2l_4v_4 \tag{6.492}$$

and

$$\begin{aligned} \lambda_i(t_f) &= 0, \qquad i = 1, 2, \ldots, 14 \\ \nu_j(t_0) = \nu_j(t_f) &= 0, \qquad j = 1, 2, \ldots, 16 \end{aligned} \tag{6.493}$$

Equations (6.370)-(6.388), (6.394), and (6.407)-(6.458) constitute a boundary value problem for global control, while equations (6.370)-(6.383), (6.389)-(6.393), and (6.459)-(6.493) constitute a boundary value problem for local control, and these equations are to be solved for the feedback parameters of the system.

Before going further, let us note that, by integrating both sides of equations (6.423)-(6.454) and using the boundary conditions (6.407) and (6.408), one obtains

$$\int_{t_0}^{t_f} (-\mu_1 y_i + 2m_i e_i)\, dt = 0 \tag{6.494}$$

$$\int_{t_0}^{t_f} (-\mu_1 y_{i+7} + 2m_{i+4} e_{i+4})\, dt = 0 \tag{6.495}$$

$$\int_{t_0}^{t_f} (-\mu_2 y_i + 2n_i f_i)\, dt = 0 \tag{6.496}$$

$$\int_{t_0}^{t_f} (-\mu_2 y_{i+7} + 2n_{i+4} f_{i+4})\, dt = 0 \tag{6.497}$$

$$\int_{t_0}^{t_f} (-\mu_3 y_i + 2p_i g_i)\, dt = 0 \tag{6.498}$$

$$\int_{t_0}^{t_f} (-\mu_3 y_{i+7} + 2p_{i+4} g_{i+4})\, dt = 0 \tag{6.499}$$

$$\int_{t_0}^{t_f} (-\mu_4 y_i + 2q_i h_i)\, dt = 0 \tag{6.500}$$

$$\int_{t_0}^{t_f} (-\mu_4 y_{i+7} + 2q_{i+4} h_{i+4})\, dt = 0 \tag{6.501}$$

where $i = 1, 2, 3, 4$. These relations are employed in Section 6.3.4.

For the special case of local control, equations (6.494)-(6.501) become

$$\int_{t_0}^{t_f} (-\mu_1 y_i + 2m_i e_i)\, dt = 0 \tag{6.502}$$

$$\int_{t_0}^{t_f} (-\mu_2 y_i + 2n_i f_i)\, dt = 0 \tag{6.503}$$

$$\int_{t_0}^{t_f} (-\mu_3 y_{i+7} + 2p_i g_i)\, dt = 0 \tag{6.504}$$

$$\int_{t_0}^{t_f} (-\mu_4 y_{i+7} + 2q_i h_i)\, dt = 0 \tag{6.505}$$

where $i = 1, 2, 3, 4$.

The following section explains the computational technique used to solve all the above optimal equations.

6.4.4. Computational Technique

The problem being solved can be reduced further to one of parameter optimization. Substituting equations (6.384)-(6.388) into equations (6.370)-(6.383), we obtain a system of differential equations for $y_1, yHg2, \ldots, y_{14}$ after the parameters $e_1, e_2, \ldots, e_8, f_1, f_2, \ldots, f_8, g_1, g_2, \ldots, g_8, h_1, h_2, \ldots, h_8$ have been determined. Let $\bar{y}_1(t), \bar{y}_2(t), \ldots, \bar{y}_{14}(t)$ be the solutions of the system of the differential equations with initial conditions

$$\bar{y}_1(t_0) = 0, \qquad i = 1, 2, \ldots, 14$$

Substituting $\{\bar{y}_i(t), i = 1, 2, \ldots, 14\}$ and equations (6.384)-(6.388) into equations (6.489)-(6.492) one has

$$\mu_1 = \lambda_7(1/T_{g1})G_2G_3 - 2l_1\left[\sum_{i=1}^{4} e_i\bar{y}_i(t) + \sum_{i=5}^{8} e_i\overline{y_{i+3}}(t)\right] \tag{6.506}$$

$$\mu_2 = \lambda_6(1/T_{e1})C_{r1} - 2l_2\left[\sum_{i=1}^{4} f_i\bar{y}_i(t) + \sum_{i=5}^{8} f_i\overline{y_{i+3}}(t)\right] \tag{6.507}$$

$$\mu_3 = \lambda_{14}(1/T_{g2})G_5G_6 - 2l_3\left[\sum_{i=1}^{4} g_i\bar{y}_i(t) + \sum_{i=5}^{8} g_i\overline{y_{i+3}}(t)\right] \tag{6.508}$$

$$\mu_4 = \lambda_{13}(1/T_{e2})C_{r2} - 2l_4\left[\sum_{i=1}^{4} h_i\bar{y}_i(t) + \sum_{i=5}^{8} h_i\overline{y_{i+3}}(t)\right] \tag{6.509}$$

Substituting equations (6.506)-(6.509) and $\{\bar{y}_i(t), i = 1, 2, \ldots, 14\}$ into equations (6.423)-(6.454) a system of differential equations for λ_1,

$\lambda_2, \ldots, \lambda_{14}$ is obtained. This system with terminal condition

$$\lambda_i(t_f) = 0, \qquad i = 1, 2, \ldots, 14$$

can be solved by integrating backward from t_f to t_0. First, let the solutions to the system of differential equations be $\overline{\lambda_1}(t), \overline{\lambda_2}(t), \ldots, \overline{\lambda_{14}}(t)$; then, the problem is reduced to selecting 32 parameters $(e_i, f_i, g_i, h_i, i = 1, 2, \ldots, 8)$, such that

$$\int_{t_0}^{t_f} \begin{bmatrix} -\overline{\mu_1}(t)\overline{y_i}(t) + 2m_i e_i \\ -\overline{\mu_2}(t)\overline{y_i}(t) + 2n_i f_i \\ -\mu_3(t)\overline{y_i}(t) + 2p_i g_i \\ -\mu_4(t)\overline{y_i}(t) + 2q_i h_i \end{bmatrix} dt = 0, \qquad i = 1, 2, 3, 4$$

$$\int_{t_0}^{t_f} \begin{bmatrix} -\overline{\mu_1}(t)\overline{y_{i+3}}(t) + 2m_i e_i \\ -\overline{\mu_2}(t)\overline{y_{i+3}}(t) + 2n_i f_i \\ -\overline{\mu_3}(t)\overline{y_{i+3}}(t) + 2p_i g_i \\ -\overline{\mu_4}(t)\overline{y_{i+3}}(t) + 2q_i h_i \end{bmatrix} dt = 0, \qquad i = 5, 6, 7, 8$$

and such that the cost functional

$$\begin{aligned} J = \int_{t_0}^{t_f} \Bigg\{ & \sum_{i=1}^{14} k_i \overline{y_i}(t)^2 + \sum_{i=1}^{8} (m_i e_i^2 + n_i f_i^2 + p_i g_i^2 + q_i h_i^2) \\ & + l_1 \left[\sum_{i=1}^{4} e_i \overline{y_i(t)} + \sum_{i=5}^{8} e_i \overline{y_{i+3}}(t) \right]^2 + l_2 \left[\sum_{i=1}^{4} f_i \overline{y_i}(t) + \sum_{i=5}^{8} f_i \overline{y_{i+3}}(t) \right]^2 \\ & + l_3 \left[\sum_{i=1}^{4} g_i \overline{y_i}(t) + \sum_{i=5}^{8} g_i \overline{y_{i+3}}(t) \right]^2 + l_4 \left[\sum_{i=1}^{4} h_i \overline{y_i}(t) h_i \overline{y_{i+3}}(t) \right]^2 \Bigg\} dt \end{aligned}$$

is minimized, where

$$\overline{\mu_1}(t) = \overline{\lambda_7(t)}(1/T_{g1})G_2G_3 - 2l_1 \left[\sum_{i=1}^{4} e_i y_i(t) + \sum_{i=5}^{8} e_i \overline{y_{i+3}}(t) \right] \tag{6.510}$$

$$\overline{\mu_2}(t) = \lambda_6(t)(1/T_{e1})C_{r1} - 2l_2 \left[\sum_{i=1}^{4} f_i \overline{y_i}(t) + \sum_{i=5}^{8} f_i \overline{y_{i+3}}(t) \right] \tag{6.511}$$

$$\overline{\mu_3}(t) = \overline{\lambda_{14}}(t)(1/T_{g2})G_5G_6 - 2l_3 \left[\sum_{i=1}^{4} g_i \overline{y_i}(t) + \sum_{i=5}^{8} g_i \overline{y_{i+3}}(t) \right] \tag{6.512}$$

$$\overline{\mu_4}(t) = \overline{\lambda_{13}}(t)(1/T_{e2})C_{r2} - 2l_4 \left[\sum_{i=1}^{4} h_i \overline{y_i}(t) + \sum_{i=5}^{8} h_i \overline{y_{i+3}}(t) \right] \tag{6.513}$$

This problem can be solved numerically by the method of iteration. Let

$$a_i(t) = \partial H / \partial e_i = \begin{cases} -\mu_1(t) y_i(t) + 2m_i e_i, & i = 1, 2, 3, 4 \\ -\mu_1(t) y_{i+3}(t) + 2m_i e_i, & i = 5, 6, 7, 8 \end{cases}$$

$$b_i(t) = \partial H/\partial f_i = \begin{cases} -\mu_2(t)y_i(t) + 2n_i f_i, & i = 1, 2, 3, 4 \\ -\mu_2(t)y_i(t) + 2n_i f_i, & i = 5, 6, 7, 8 \end{cases}$$

$$c_i(t) = \partial H/\partial g_i = \begin{cases} -\mu_3(t)y_i(t) + 2p_i g_i, & i = 1, 2, 3, 4 \\ -\mu_3(t)y_{i+3}(t) + 2p_i g_i, & i = 5, 6, 7, 8 \end{cases}$$

$$d_i(t) = \partial h/\partial h_i = \begin{cases} -\mu_4(t)y_i(t) + 2q_i h_i, & i = 1, 2, 3, 4 \\ -\mu_4(t)y_{i+3}(t) + 2q_i h_i, & i = 5, 6, 7, 8 \end{cases}$$

and

$$a_i = \int_{t_0}^{t_f} a_i(t)\, dt \tag{6.514}$$

$$b_i = \int_{t_0}^{t_f} b_i(t)\, dt \tag{6.515}$$

$$c_i = \int_{t_0}^{t_f} c_i(t)\, dt \tag{6.516}$$

$$d_i = \int_{t_0}^{t_f} d_i(t)\, dt \tag{6.517}$$

where $i = 1, 2, \ldots, 8$ and where

$$\mu_1(t) = \lambda_7(t)(1/T_{g1})G_2G_3 - 2l_1\left[\sum_{i=1}^{4} e_i y_i(t) + \sum_{i=5}^{8} e_i y_{i+3}(t)\right] \tag{6.518}$$

$$\mu_2(t) = \lambda_6(t)(1/T_{e1})C_{r1} - 2l_2\left[\sum_{i=1}^{4} f_i y_i(t) + \sum_{i=5}^{8} f_i y_{i+3}(t)\right] \tag{6.519}$$

$$\mu_3(t) = \lambda_{14}(t)(1/T_{g2})G_5G_6 - 2l_3\left[\sum_{i=1}^{4} g_i y_i(t) + \sum_{i=5}^{8} g_i y_{i+3}(t)\right] \tag{6.520}$$

$$\mu_4(t) = \lambda_{13}(t)(1/T_{e2})C_{r2} - 2l_4\left[\sum_{i=1}^{4} h_i y_i(t) + \sum_{i=5}^{8} h_i y_{i+3}(t)\right] \tag{6.521}$$

In the above relations, $\{y_i(t),\ i = 1, 2, \ldots, 14\}$ and $\{\lambda_i(t),\ i = 1, 2, \ldots, 14\}$ are the solutions of the systems of equations (6.370)–(6.383) and (6.459)–(6.472) for a preselected set of parameters e_i, f_i, g_i, h_i, $i = 1, 2, \ldots, 8$.

The iteration procedure is given by deduction.

The first guess of the control parameters $\{e_i^0, f_i^0, g_i^0, h_i^0, i = 1, 2, \ldots, 8\}$ is arbitrarily assigned. If the control parameters of the jth iteration,

$$\{e_i^j, f_i^j, g_i^j, h_i^j, i = 1, 2, \ldots, 8\}$$

are determined, then the ones of the $(j + 1)$th iteration,

$$\{e_i^{j+1}, f_i^{j+1}, g_i^{j+1}, h_i^{j+1}, i = 1, 2, \ldots, 8\}$$

take the form

$$e_i^{j+1} = e_i^j - \alpha_i^j a_i^j \tag{6.522}$$

$$f_i^{j+1} = f_i^j - \beta_i^j b_i^j \tag{6.523}$$

$$g_i^{j+1} = g_i^j - \gamma_i^j c_i^j \tag{6.524}$$

$$h_i^{j+1} = h_i^j - \delta_i^j d_i^j \tag{6.525}$$

where $i = 1, 2, \ldots, 8$ and where the α_i^j's, β_i^j's, γ_i^j's, δ_i^j's are suitably selected constants. The constants a_i^j's, b_i^j's, c_i^j's, d_i^j's, which also depend on the iteration number i, are found from equations (6.514)–(6.517) by using the control parameters of the jth iteration.

The stopping criteria are given by

$$|a_i| < \varepsilon \tag{6.526}$$

$$|b_i| < \varepsilon \tag{6.527}$$

$$|c_i| < \varepsilon \tag{6.528}$$

$$|d_i| < \varepsilon \tag{6.529}$$

for all i, $i = 1, 2, \ldots, 8$, and the cost functional is to be kept reducing with increasing j. Conditions (6.526)–(6.529) are selected as stopping criteria, because numerically speaking they are equivalent to (6.494)–(6.501) if ε is a small, positive number.

In practical computations, all of the α_i^j's, β_i^j's, γ_i^j's, δ_i^j's are taken to be equal to, say, ω^j, which is used to obtain the control parameters of the $(j+1)$th iteration from the ones of the jth iteration. It is noticed that ω^j

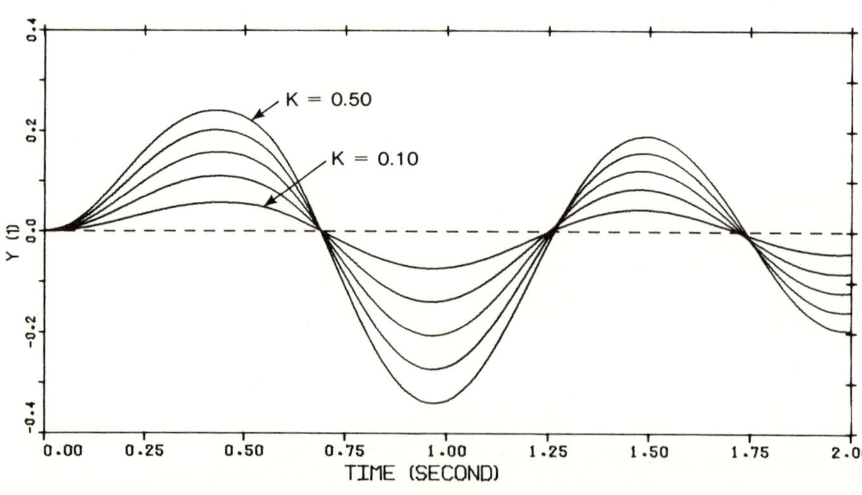

Figure 6.16. State variable $Y(1)$ versus time.

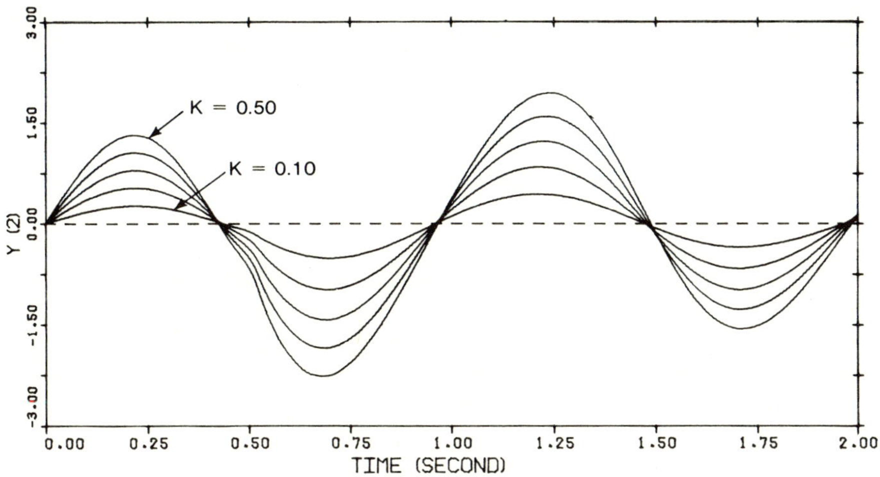

Figure 6.17. State variable $Y(2)$ versus time.

varies with the number of iterations j for the fixed strength of torque pulse K.

ω^j is determined as follows. Suppose that the control parameters of the jth iteration, for which the stopping criteria are not met, are obtained. All the a_i^j's, b_i^j's, c_i^j's, d_i^j's are then obtained from equations (6.510)–(6.517) by using the control parameters of the jth iteration. For a given $\bar{\omega}$, letting

$$\alpha_i^j = \beta_i^j = \gamma_i^j = \delta_i^j = \bar{\omega}$$

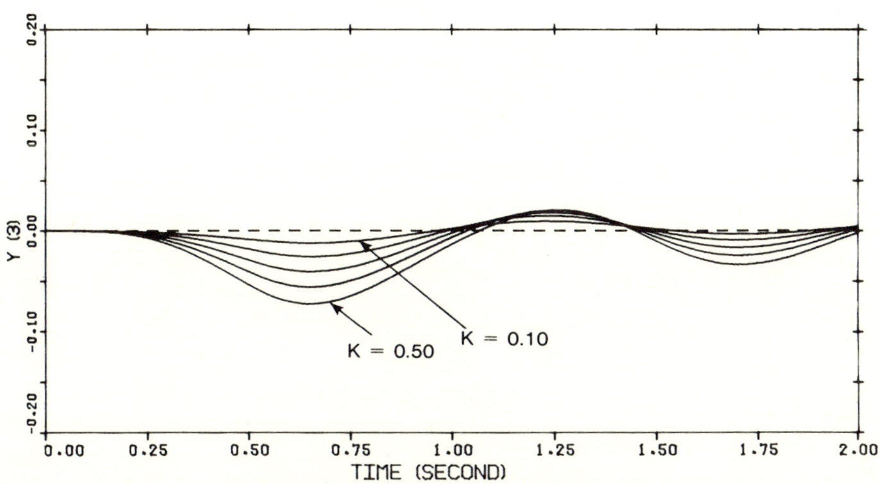

Figure 6.18. State variable $Y(3)$ versus time.

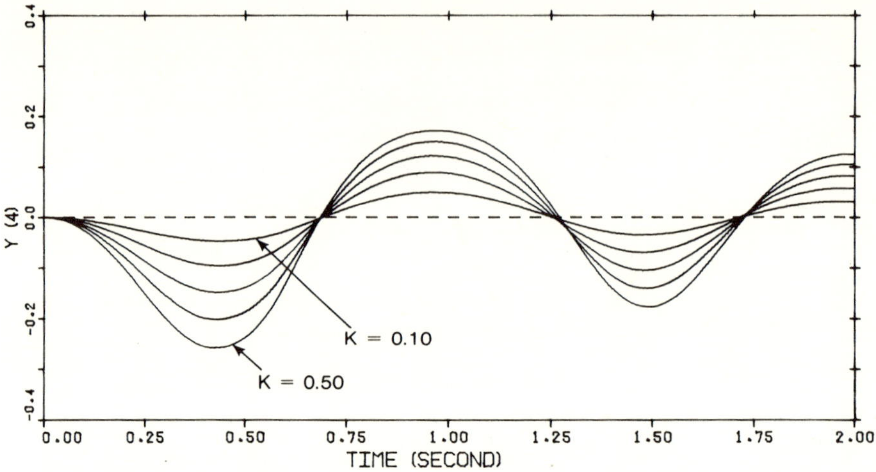

Figure 6.19. State variable $Y(4)$ versus time.

one obtains

$$\bar{e}_i = e_i^j - \bar{\omega} a_i^j \tag{6.530}$$

$$\bar{f}_i = f_i^j - \bar{\omega} b_i^j \tag{6.531}$$

$$\bar{g}_i = g_i^j - \bar{\omega} c_i^j \tag{6.532}$$

$$\bar{h}_i = h_i^j - \bar{\omega} d_i^j \tag{6.533}$$

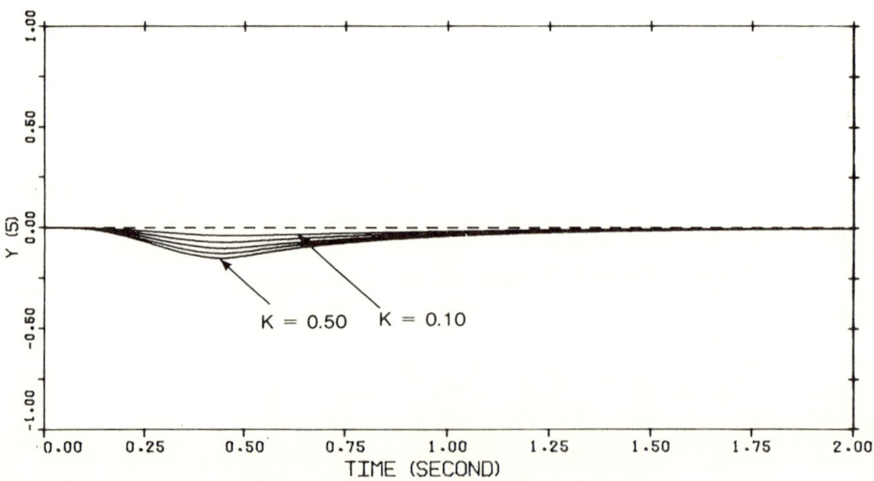

Figure 6.20. State variable $Y(5)$ versus time.

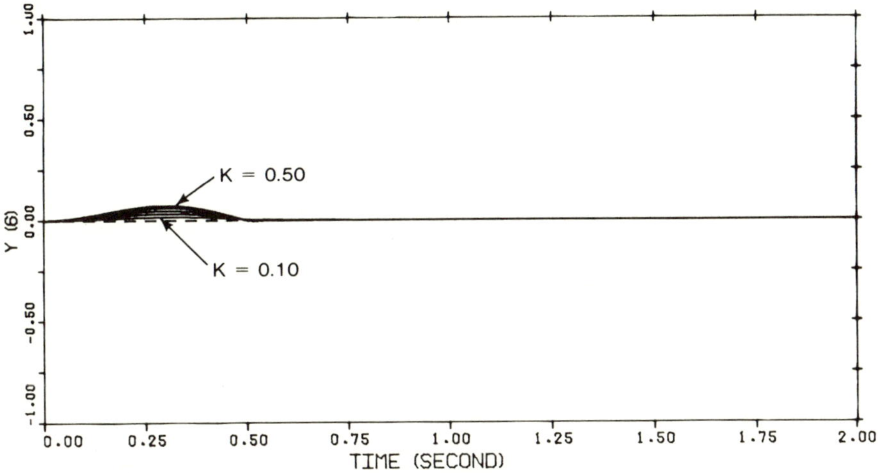

Figure 6.21. State variable $Y(6)$ versus time.

where $i = 1, 2, \ldots, 8$. Substituting

$$v_1 = \sum_{i=1}^{4} \bar{e}_i y_i + \sum_{i=5}^{8} \bar{e}_i y_{i+3}$$

$$v_2 = \sum_{i=1}^{4} \bar{f}_i y_i + \sum_{i=5}^{8} \bar{f}_i y_{i+3}$$

$$v_3 = \sum_{i=1}^{4} \bar{g}_i y_i + \sum_{i=5}^{8} \bar{g}_i y_{i+3}$$

$$v_4 = \sum_{i=1}^{4} \bar{h}_i y_i + \sum_{i=5}^{8} \bar{h}_i y_{i+3}$$

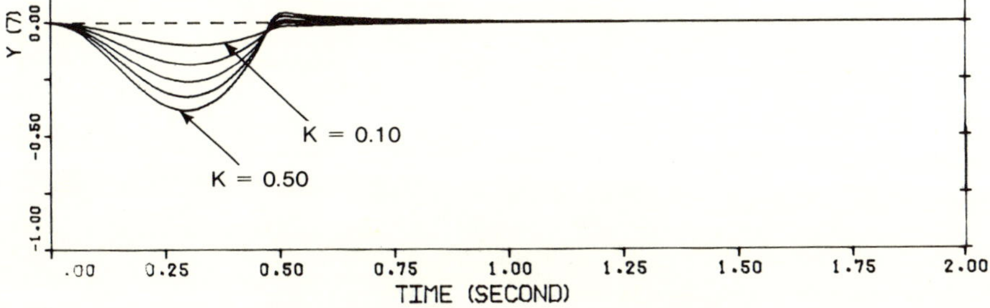

Figure 6.22. State variable $Y(7)$ versus time.

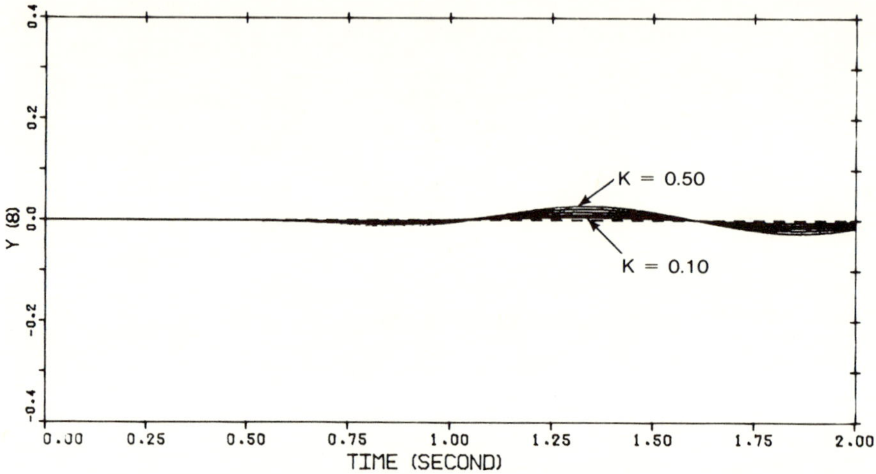

Figure 6.23. State variable $Y(8)$ versus time.

into equations (6.370)–(6.383) with the help of the initial conditions (6.394) we obtain the solutions for the differential equation (6.370)–(6.383). After further substitution of these solutions into equations (6.369) the cost functional $\bar{J}$ can be obtained by integration. At this point, it is noted that $\bar{J}$ is a function of $\bar{\omega}$. ω^j is selected such that $\bar{J}$ is minimized. In so doing, the cost functional can always be reduced.

The differential equations (6.370)–(6.383) and (6.459)–(6.472) are solved numerically forward and backward, respectively, by fourth-order Runge–Kutta algorithms with step size $k = 0.005$. This step size is good

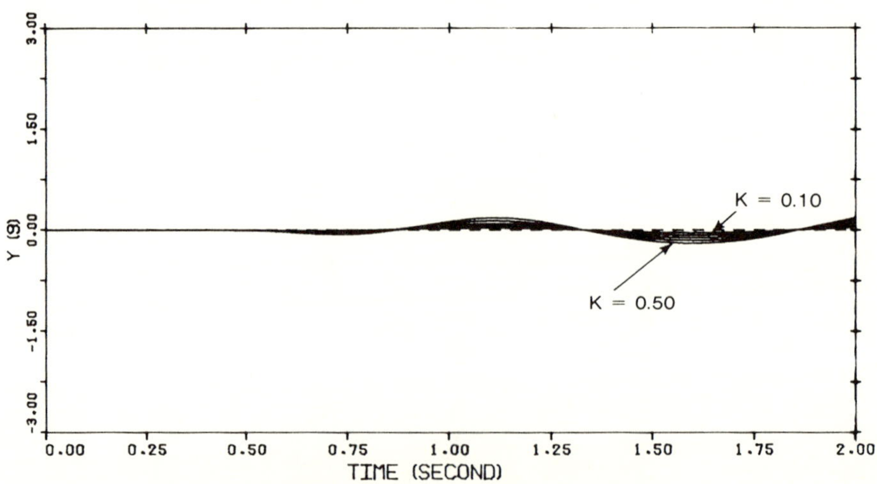

Figure 6.24. State variable $Y(9)$ versus time.

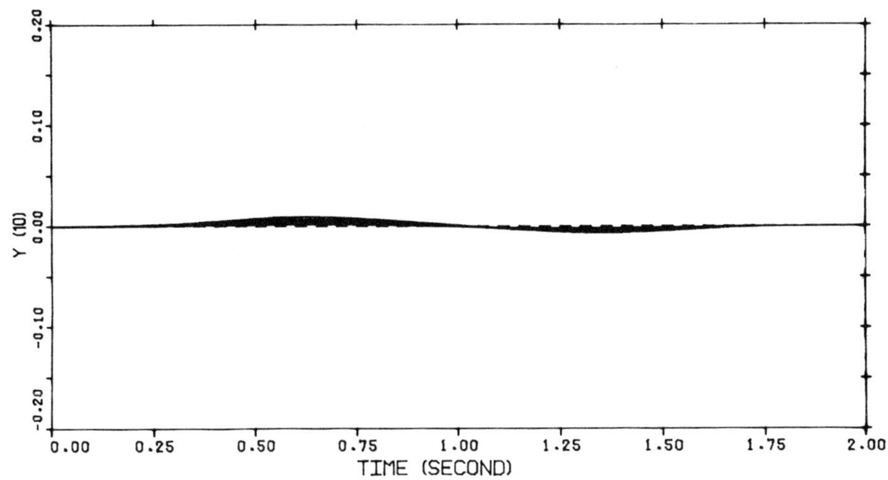

Figure 6.25. State variable $Y(10)$ versus time.

enough for the practical cases ($K = 0.0$ to $K = 0.50$); but, for large disturbances (such as $K = 2.0$), a smaller step size is required to produce accurate results.

The optimal solutions are obtained numerically from $t_0 = 0.0$ to $t_f = 0.50$ for both global and local controls. For testing purposes, the optimal solutions for two separated machine systems (in which case, the coupling between the two machines is broken) are also obtained.

The optimal parameters for the case $K = 0.0$, all of which are zeros, are obtained first. These values are then used as a first guess to find the

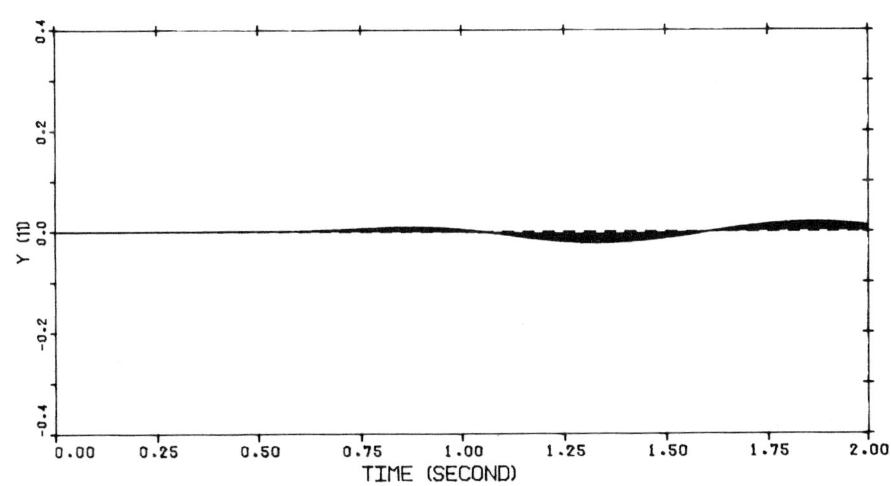

Figure 6.26. State variable $Y(11)$ versus time.

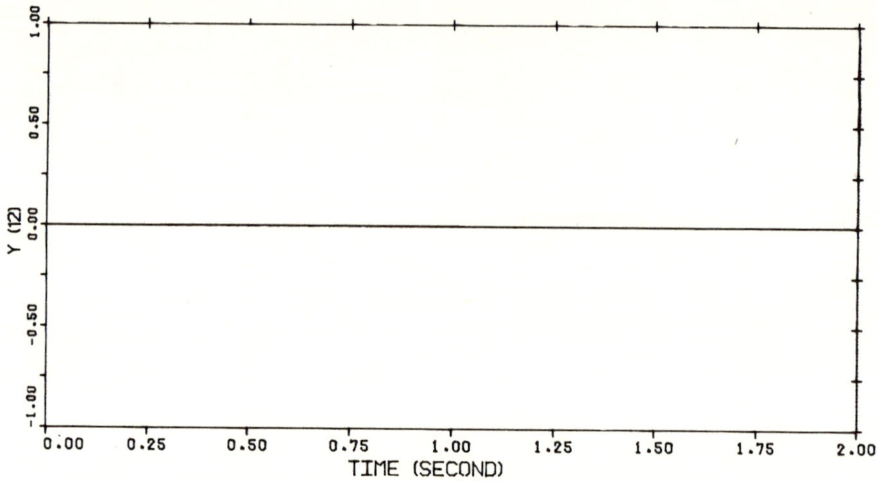

Figure 6.27. State variable $Y(12)$ versus time.

optimal parameters for the case $K = 0.05$. For the succeeding search for the optimal parameters ($K \geq 0.10$), the same procedure is applied. This method usually provides a good initial guess, because of the continuity of the optimal parameters with respect to K, and it was found that the rate of convergence is considerable.

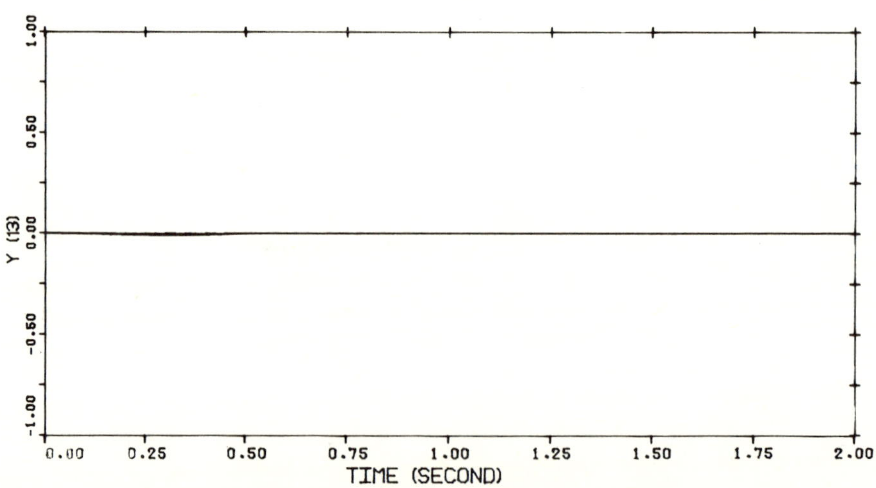

Figure 6.28. State variable $Y(13)$ versus time.

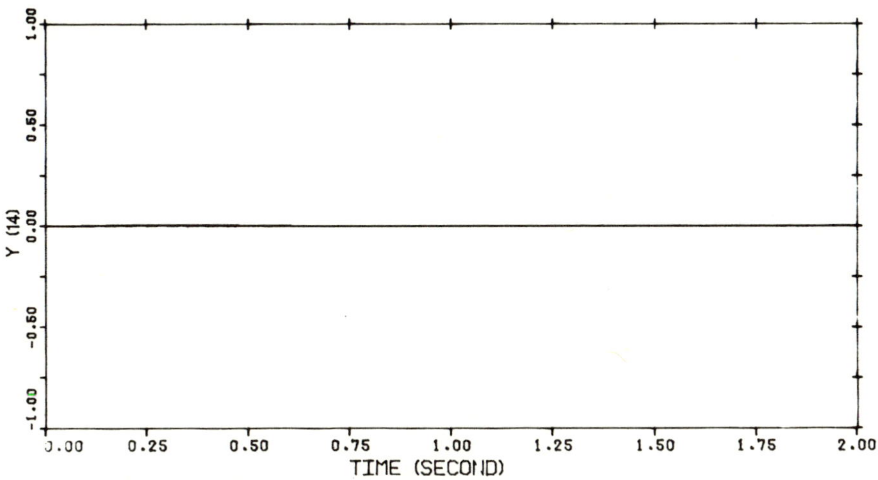

Figure 6.29. State variable $Y(14)$ versus time.

The results are shown in Figs. 6.16–6.33. The results are similar to the ones obtained in Section 6.2. Inclusion of the turbine, governor, and exciter time constants does, however, give more time lag than one would expect in damping out the disturbance. It is clear from the figures that applying a torque pulse disturbance to machine number 1 has little effect on the behavior of machine number 2 when both machines carry a normal load, which is the case considered here.

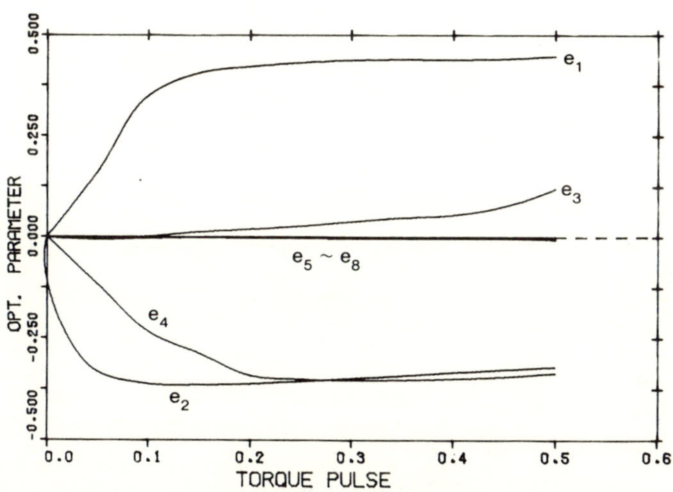

Figure 6.30. Feedback parameters versus torque pulse.

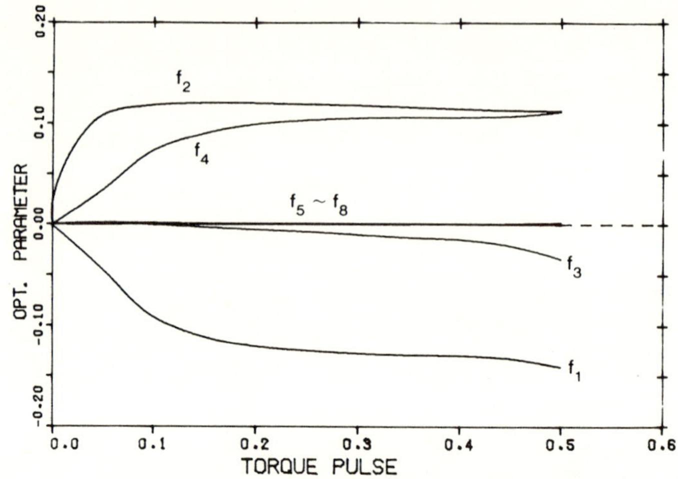

Figure 6.31. Feedback parameters versus torque pulse.

The effect of changing the coupling impedance, x_3, from $x_3 = 0.10$ to $x_3 = 0.01$ has also been computed. It is noted that such a change has negligible effect on the optimal parameters and the state variables.

6.5. Conclusions

In this chapter the problems of the optimal control of turboalternators are discussed. In the first section, the optimal control of a large turboalternator has been investigated. All quantities have been put in per unit values;

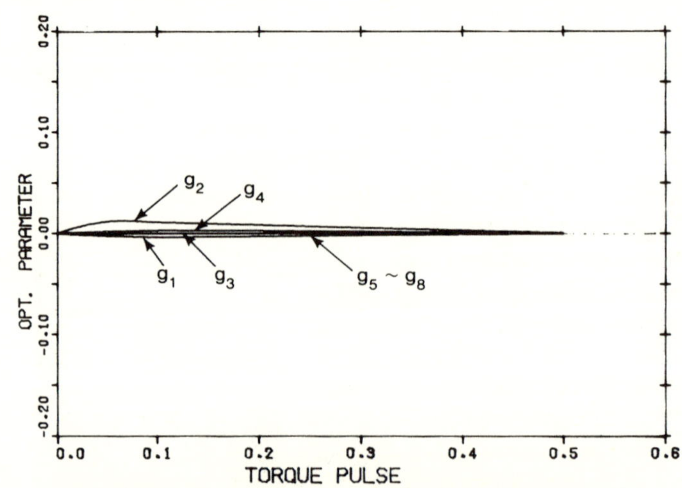

Figure 6.32. Feedback parameters versus torque pulse.

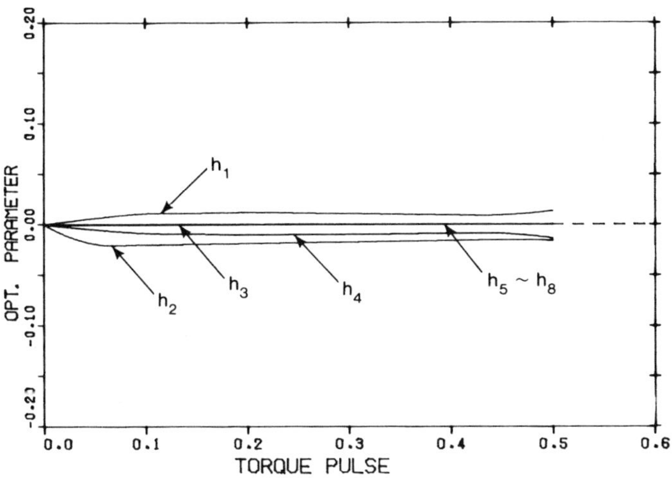

Figure 6.33. Feedback parameters versus torque pulse.

hence, the problem is general in nature. It should be noted that the nonlinear transformation used for casting the problem into quadratic form is quite general for stability studies in power systems. The problem posed here is highly ill conditioned, and this situation was remedied by using the Riccati equation to get an estimate for the initial values of the Lagrange multipliers of the nonlinear system. As noted, a very satisfactory solution was obtained by this method.

In the second and third sections, the optimizing equations for two interconnected turbogenerators are obtained. These equations become the ones investigated in the first section, when the coupling between the two machines is broken. The optimal feedback parameters are obtained numerically by applying the gradient method to the nonlinear two-point boundary value problem. The values obtained for these parameters depend on the strength K and the duration τ of the disturbance, since the model is nonlinear, contrary to the usual feedback control of a linear model. Using the optimal feedback parameters, the nonlinear differential equations for the state variables are integrated by the Runge–Kutta method forward from $t_0 = 0.0$ to $t_f = 0.5$ s, and then the nonlinear costate equations are integrated by using the same method backward from $t_f = 0.5$ to $t_0 = 0.0$ s. The value of the cost functional is obtained by using Simpson's method.

Both global and local controls are considered. The numerical results indicate that there is not much difference between these two kinds of controls. This outcome will simplify the design of linear feedback controllers. Furthermore, the numerical solutions also indicate that the two-machine system can withstand a larger disturbance than the one-machine system if no controls are executed.

To see how the coupling impedance $z_3 = jx_3$ affects the two-interconnected-machine system, the optimal parameters and solutions to the system of optimality equations are obtained for the cases $x_3 = 0.1$ and $x_3 = 0.01$ (see Tables 6.3–6.6). Further computations indicate that the optimal solutions for the cases $x_3 \leq 0.10$ do not differ much from those shown in this section.

Appendix 6.A

Equations (6.1)–(6.3) are derived in the following manner. The voltages in the quadratic and direct axes are given in Ref. 6.5 as

$$u_d = p\Psi_d + \omega\Psi_q + R_a i_d \tag{6.A.1}$$

$$u_q = p\Psi_q - \omega\Psi_d + R_a i_d \tag{6.A.2}$$

$$u_f = p\Psi_f^1 + R_f i_f \tag{6.A.3}$$

Here, (6.A.1) and (6.A.2) hold for the stator, and (6.A.3) holds for the rotor. The flux linkage equations are

$$\begin{aligned} \omega_0\Psi_d &= x_d i_d + x_{md} i_f \\ \omega_0\Psi_q &= x_q i_q \\ \omega_0\Psi_f^1 &= x_{md} i_d + x_{fd} i_f \end{aligned} \tag{6.A.4}$$

The torque equation is

$$M_t = M_e - (2H/\omega_0)(d^2\delta^1/dt^2) - K_d(d\delta^1/dt) \tag{6.A.5}$$

It is known that the machine is connected to an infinite bus by a series impedance x_e, R_e. Then, one can change

$$\begin{aligned} x_d &\to x_d^1, \qquad \text{where } x_d^1 = x_d + x_e \\ x_q &\to x_q^1, \qquad \text{where } x_q^1 = x_q + x_e \\ R_a &\to R_a^1, \qquad \text{where } R_a^1 = R_a + R_e \end{aligned} \tag{6.A.6}$$

and with the primed quantities one can then consider the machine connected directly to the infinite bus.

It is assumed that

$$p\Psi_d \sim 0, \qquad p\Psi_q \sim 0, \qquad R_a \sim 0, \qquad R_e \sim 0, \qquad \omega \sim \omega_0 \tag{6.A.7}$$

in the voltage equations. It is desired to adopt the convention that the machine acts as a generator. Therefore, the definition of the angle is changed to

$$\delta = -\delta^1$$

One also defines

$$\Psi_f = -\Psi_f^1$$

Equations (6.A.1)-(6.A.5) now become

$$u_d = \omega_0 \Psi_q \tag{6.A.8}$$

$$u_q = -\omega_0 \Psi_d \tag{6.A.9}$$

$$\omega_0 \Psi_d = x_d^1 i_d + x_{md} i_f \tag{6.A.10}$$

$$\omega_0 \Psi_q = x_q^1 i_q \tag{6.A.11}$$

$$\omega_0 \Psi_f = -x_{md} i_d - x_{fd} i_f \tag{6.A.12}$$

$$d^2\delta/dt^2 = (\omega_0/2H)[M_t - M_e - K_d(d\delta/dt)] \tag{6.A.13}$$

and

$$M_e = (\omega_0/2)(\Psi_d i_q - \Psi_q i_d) \tag{6.A.14}$$

One also has

$$(d/dt)(\omega_0 \Psi_f) = -\omega_0 u_f + \omega_0 R_f i_f \tag{6.A.15}$$

The bus phase voltage is

$$u_a = u_{\text{rms}}\sqrt{2}\sin(\omega_0 t) = u_m \sin(\omega_0 t) \tag{6.A.16}$$

This is related to the quadrature and direct axes by Park's transformation, where

$$u_a = u_d \cos\theta + u_q \sin\theta \tag{6.A.17}$$

and

$$\theta = \omega_0 t + \delta$$

Now, comparing (6.A.15) and (6.A.17), one obtains

$$u_d = -u_m \sin\delta, \qquad u_q = u_m \cos\delta \tag{6.A.18}$$

Eliminating the currents i_d, i_q, i_f, one obtains the machine equations

$$d^2\delta/dt^2 = (1/M)[M_t - s_4 \sin\delta\cos\delta - s_5\omega_0\Psi_f \sin\delta - K_d(d\delta/dt)]$$

$$d(\omega_0\Psi_f)/dt = -\omega_0 u_f - A\omega_0\Psi_f + c\cos\delta \tag{6.A.19}$$

where

$$M = 2H/\omega_0$$

and

$$s_4 = \frac{-u_{\text{rms}}^2(x_q^1 - x_d^1 + x_{md}^2/x_{fd})}{x_q^1(x_d^1 - x_{md}^2/x_{fd})}$$

$$s_5 = (u_{\text{rms}}/\sqrt{2})(x_{md}/x_{fd})\left(\frac{1}{x_d^1 - x_{md}^2/x_{fd}}\right)$$

$$A = (x_d^1/x_{fd})\left(\frac{1}{x_d^1 - x_{md}^2/x_{fd}}\right)\omega_0 R_f$$

$$c = \frac{\sqrt{2}\,u_{\text{rms}}\omega_0 R_f x_{md}}{x_{fd}(x_d^1 - x_{md}^2/x_{fd})}$$

Appendix 6.B: Steady-State Values and Machine Constants (Refs. 6.1 and 6.6)

Generator rating = 37.5 MVA or 30 MW at 0.8 p.f. at 11.8 kV line-to-line
Number of poles = 2; rpm = 3000, $f = 50$ Hz
Infinite bus voltage = 1.00 per unit per phase
$H = 6.63$ MW s/MVA
$K_d = 0.02535$
$R_f = 0.00107$ per unit
$x_{md} = 1.86$ per unit
$x_{mq} = 1.86$ per unit
$x_a = 0.14$ per unit (leakage reactance)
$x_{f1} = 0.14$ per unit (leakage reactance)
$x_d = x_{md} + x_a + x_e = 2.00 + x_e$ per unit
$x_q^1 = x_{mq} + x_a + x_e = 2.00 + x_e$ per unit
$x_{fd} = x_{md} + x_f = 2.00$ per unit
$x_e = 0.181$ per unit

From these now, with the circuit breaker closed,

$$s_4 = -1.76, \qquad s_5 = 1.46, \qquad A = 1.812, \qquad c = 0.980$$

During the fault,

$$s_4 = s_5 = c = 0, \qquad A = 0.170$$

Steady-state value of flux $\omega_0 \Psi_f^s = 1.874$ per unit
Steady-state value of torque $M_t^s = 0.800$ per unit
Steady-state value of torque angle $\delta^s = 37.1°$, and $d\delta^s/dt = 0$
Steady-state value of field voltage $-\omega_0 u_f = 0.738$

By definition,

$$u_1 = M_t, \qquad u_2 = -\omega_0 u_f$$

For the cost functional penalty factor, one takes

$$\alpha = 2.50, \quad \alpha_2 = 1.00, \quad \alpha_3 = 0.100, \quad \beta_1 = 1.00, \quad \beta_2 = 1.00$$

Appendix 6.C: Steady-State Values and Machine Constants

It is assumed that the two machines are identical.
Generator rating = 37.5 MVA or 30 MW at 0.8 p.f. at 11.8 kV, line-to-line
Number of poles = 2; rpm = 3000, $f = 50$ Hz
Infinite bus voltage = 1.00 per unit per phase
$H_1 = H_2 = 6.63$ MW s/MVA
$K_{d1} = K_{d2} = 0.02535$
$R_{f1} = R_{f2} = 0.00107$ per unit
$x_{md1} = x_{md2} = 1.86$ per unit
$x_{mq1} = x_{mq2} = 1.86$ per unit
$x_{a1} = x_{a2} = 0.14$ per unit (leakage reactance)
$x_{f1} = x_{f2} = 0.14$ per unit (leakage reactance)
$x_{d1} = x_{md1} = 2.00$ per unit
$x_{d2} = x_{md2} + x_{a2} = 2.00$ per unit
$x_{q1} = x_{mq1} + x_{a1} = 2.00$ per unit
$x_{q2} = x_{mq2} + x_{a2} = 2.00$ per unit
$x_{fd1} = x_{md1} + x_{f1} = 2.00$ per unit
$x_{fd2} = x_{md2} + x_{f2} = 2.00$ per unit
$x_1 = 0.181$ per unit, $x_2 = 0.181$ per unit, $x_3 = 0.10$ per unit

For these values, we now have

$R_1^* = -1.12396, \quad R_2^* = -0.17289$
$S_1^* = 1.0085, \quad S_2^* = -0.15396$
$S_3^* = 0.24542, \quad S_4^* = 0.12765$
$R_1 = -0.17289, \quad R_2 = -1.12396$
$S_1 = -0.15396, \quad S_2 = 1.00085$
$S_1 = -0.15396, \quad S_2 = 1.00085$
$S_3 = 0.12765, \quad S_4 = 0.24542$
$a_1 = -37.91885, \quad b_1 = 40.77296$
$c_1 = 1.15699, \quad d_1 = 1.24408$
$a_2 = -37.91885, \quad b_2 = 40.77296$
$c_2 = 1.15699, \quad d_2 = 1.24408$
$u_{d1}^s = -0.78193, \quad u_{q1}^s = 1.37422$
$u_{d2}^s = -0.78193, \quad u_{q2}^s = 1.37422$

Steady-state value of torque angle $\omega_1^s = \delta_2^s = 37.0814°$
Steady-state value of flux $\omega_0\psi_{f1}^s = \omega_0\psi_{f2}^s = 1.87253$ per unit
Steady-state value of torque $M_{t1}^s = M_{t2}^s = 0.8$ per unit
Steady-state value of field voltage $-\omega_0 u_{f1} = -\omega_0 u_{f2} = 0.73962$ per unit

By definition,

$$
\begin{aligned}
&x_1^s = \sin \delta_1^s = 0.60295, \qquad x_2^s = 0 \\
&x_3^s = \omega_0\psi_{f1}^s = 1.87253, \qquad x_4^s = \cos \delta_1^s = 0.79778 \\
&x_5^s = \delta_2^s = 0.60295, \qquad x_6^s = 0 \\
&x_7^s = \omega_0\psi_{f2}^s = 1.87253, \qquad x_8^s = \cos \delta_2^s = 0.79778 \\
&U_1^s = M_{t1}^s = 0.8 \text{ p.u.}, \qquad V_1^s = -\omega_0 u_{f1}^s = 0.73962 \text{ p.u.} \\
&U_2^s = M_{t2}^s = 0.8 \text{ p.u.}, \qquad V_2^s = -\omega_0 u_{f2}^s = 0.73962 \text{ p.u.}
\end{aligned}
$$

For the cost functional penalty factor, one takes

$$k_1 = k_4 = k_5 = k_8 = 2.50$$

$$k_2 = k_6 = 1.0$$

$$k_3 = k_7 = 0.10$$

$$l_1 = l_2 = l_3 = l_4 = 1.0$$

$$m_i = n_i = p_i = q_i = 0.001, \qquad i = 1, 2, \ldots, 8$$

Appendix 6.D: Steady-State Values and Machine Constants

It is assumed that the two machines are identical (Ref. 6.1). The generators' rating is 37.5 MVA or 30 MW at 0.8 p.f. at 11.8 kV, line-to-line.

The number of poles is 2; rpm = 3000; $f = 50$ Hz. The infinite bus voltage is 1.00 per unit per phase. The machine constants are

$H_1 = H_2 = 6.63$ MW s/MVA
$K_{d1} = K_{d2} = 0.02535$
$R_{f1} = R_{f2} = 0.00107$ per unit
$x_{md1} = x_{md2} = 1.86$ per unit
$x_{mq1} = x_{mq2} = 1.86$ per unit
$x_{a1} = x_{a2} = 0.14$ per unit (leakage reactance)
$x_{f1} = x_{f2} = 0.14$ per unit (leakage reactance)
$x_{d1} = x_{md1} + x_{a1} = 2.00$ per unit
$x_{d2} = x_{md2} + x_{a2} = 2.00$ per unit
$x_{q1} = x_{mq1} + x_{a1} = 2.00$ per unit
$x_{q2} = x_{mq2} + x_{a2} = 2.00$ per unit
$x_{fd1} = x_{md2} + x_{f1} = 2.00$ per unit
$x_{fd2} = x_{md2} = x_{f2} = 2.00$ per unit
$x_1 = 0.181$ per unit
$x_2 = 0.181$ per unit

$x_3 = 0.10$ per unit
$G_2 = G_5 = 1.33$
$G_3 = G_6 = 1.42$
$G_1 = G_4 = 0.00188$ (neglected)
$T_{g1} = T_{g2} = 0.2$
$T_{b1} = T_{b2} = 0.49$
$T_{e1} = T_{e2} = 0.2$
$C_{r1} = C_{r2} = 1.0$

For these values, we see that

$R_1^* = R_2 = -1.12396$
$R_2^* = R_1 = -0.17289$
$S_1^* = S_2 = 1.00085$
$S_2^* = S_1 = -0.15396$
$S_3^* = S_4 = 0.24542$
$S_4^* = S_3 = 0.12765$
$a_1 = a_2 = -37.91885$
$b_1 = b_2 = 4.077296$
$c_1 = c_2 = 1.15699$
$d_1 = d_2 = 1.24408$
$u_{d1}^s = u_{d2}^s = -0.78193$
$u_{q1}^s = u_{q2}^s = 1.37422$

The steady-state value of the torque angle is

$$\delta_1^s = \delta_2^s = 37.0814°$$

The steady-state value of the flux is

$$\omega_0 \psi_{fd1}^s = \omega_0 \psi_{fd2}^s = 1.87253 \text{ p.u.}$$

The steady-state value of the torque is

$$M_{t1}^s = x_5^s = x^s = M_{t2}^s = 0.8 \text{ p.u.}$$

The steady-state value of the field voltage is

$$\omega_0 u_{f1}^s = x_6^s = x_{13}^s = \omega_0 u_{f2}^s = 0.73962 \text{ p.u.}$$

By definition,

$x_1^s = \sin \delta_1^s = 0.60285$
$x_2^s = 0$
$x_3^s = \omega_0 \psi_{fd1}^s = 1.87253$
$x_4^s = \cos \delta_1^s = 0.79778$
$x_8^s = \sin \delta_2^s = 0.60295$
$x_9^s = 0$
$x_{10}^s = \omega_0 \psi_{fd2}^s = 1.87253$
$x_{11}^s = \cos \delta_2^s = 0.79778$

For the cost functional penalty factors, one takes

$$k_1 = k_4 = k_8 = k_{11} = 2.50$$

$$k_2 = k_9 = 1.0$$

$$k_3 = k_{10} = 0.10$$

$$k_5 = k_6 = k_7 = k_{12} = k_{13} = k_{14} = 0.10$$

$$l_1 = l_2 = l_3 = l_4 = 1.0$$

$$m_i = n_i = p_i = q_i = 0.001, \qquad i = 1, 2, \ldots, 8$$

References

6.1. IYER, S. N., and CORY, B. J., "Optimization of Turbogenerator Transient Performance by Differential Dynamic Programming," *IEEE Trans. Power Appar. Syst.* **PAS-90**, 2149–2157 (1971).

6.2. YU, Y., VONGSURIYA, K., and WEDMAN, L. N., "Application of an Optimal Control Theory to a Power System," *IEEE Trans. Power Appar. Syst.* **PAS-89**, 55–62 (1970).

6.3. SHAMALY, A., CHRISTENSEN, G. S., and EL-HAWARY, M. E., "Functional Optimization of Systems with Polynomial Nonlinearities," Proceedings of the 20th Midwest Symposium on Circuits and Systems, Lubbock, Texas, 1977.

6.4. PESCHON, J., PIERCY, D. S., TINNEY, W. F., TVEIT, O. J., and CUENOD, M., "Optimum Control of Reactive Power Flow," *IEEE Trans. Power Appar. Syst.* **PAS-87**, 44–48 (1968).

6.5. ADKINS, B., and HARLEY, G. R., *The General Theory of Alternating Current Machines*, Chapman and Hall, London, England, 1975.

6.6. SHACKSHAFT, G., "General Purpose Turboalternator Model," *IEE Proc.* **110**, 703–713 (1963).

6.7. SHAMALY, A., CHRISTENSEN, G. S., and EL-HAWARY, M. E., "Optimal Control of a Large Turboalternator," *J. Opt. Theory Appl.* **34**, 83–97 (1971).

6.8. SHAMALY, A., CHRISTENSEN, G. S., and EL-HAWARY, M. E., "A Transformation for Necessary Optimality Conditions for Systems with Polynomial Nonlinearities," *IEEE Trans. Autom. Control* **AC-24**, 983–984 (1979).

6.9. ATHANS, M., and FALB, P. L., *Optimal Control*, McGraw-Hill, New York, 1966.

6.10. SAGE, A. P., and WHITE, C. C., *Optimum Systems Control*, Prentice-Hall, Englewood Cliffs, New Jersey, 1977.

6.11. DAVISON, E. J., RAU, N. S., and PALMAY, F. V., "The Optimal Decentralized Control of a Power System Consisting of a Number of Interconnected Synchronous Machines," *Int. J. Control* **18**, 1313–1323 (1973).

6.12. SHAMALY, A., CHRISTENSEN, G. S., and EL-HAWARY, M. E., "Realistic Feedback Control of Turbogenerators," *J. Opt. Theory Appl.* **35**(2), 251–259 (1981).

6.13. IYER, S. N., and CORY, B. J., "Optimal Control of a Turbogenerator Including an Exciter and Governor," *IEEE Trans. Power Appar. Syst.* **PAS-90**, 2142–2148 (1971).

6.14. SHAMALY, A., CHRISTENSEN, G. S., and CHEN, Y., "Optimal Control of Two Interconnected Turbogenerators," *J. Opt. Theory Appl.* **40**(2), 255–291 (1983).

6.15. CHEN, Y., CHRISTENSEN, G. S., and SHAMALY, A., "Realistic Feedback Control of Two Interconnected Turbogenerators," *J. Opt. Theory Appl.* **42**(1), 103–141 (1984).

7

Conclusion

7.1. Summary

A range of problems concerning optimal control applications in electric power systems has been considered in the previous chapters. It is clear that such problems can only now be solved numerically, because of the advent of large-scale digital computers.

Following the introductory background chapter we considered dynamic optimal load flow in Chapter 3. Here we assumed that hydroplants have constant head and developed two iterative methods for the solution of the optimal hydrothermal load flow problem. Several examples were solved numerically. In Chapter 4 we considered economic dispatch of hydrothermal–nuclear systems and we concluded this chapter by presenting a two-level iterative scheme suitable for numerical solutions of the above problem. Chapter 5 focused on the problem of optimal tie-line control. Pontryagin's maximum principle was the tool employed in that chapter for designing regulators for the problem of load frequency control of interconnected power systems. Chapter 6 treated problems in the area of optimal torque and voltage control of turboalternators. Three different problems were considered: (1) A turboalternator connected to an infinite bus, (2) two interconnected identical turboalternators connected to an infinite bus, and (3) two interconnected identical turboalternators connected to an infinite bus with inclusion of equipment time constants.

7.2. Future Work

In looking toward future research needs it is clear that an algorithm is needed for optimal hydrothermal load flow with variable head. Many hydroplants exhibit significant head variations during the optimization period. A more general formulation of the hydrothermal–nuclear problem is desirable, for example, with chains of hydroplants on different rivers, as is often the case in practice. In this book we really only gave an introduction to the problem of optimal torque and voltage control of turboalternators. In this area too there is room for much further research, for example, into the problem of differently rated alternators with different loadings.

Index

Active and reactive power, 23
Active power balance equation, 9, 17, 23
Active power balance model, 22
Adjoint conjugate operator, 18
Admissible control, 10
All-thermal electric power system, 21, 22
All-thermal problem, 2
Angular frequency deviation, 100
Area bias setting, 90
Area control error (ace), 72, 82
Augmented cost functional, 9

Backlash element, 3, 103
Backward integration, 118
Banach space, 18
Bellman, 1
Bias-setting, 82
Boundary conditions, 8
Boundary part, 115
Bus phase voltage, 185

Calculus of variations, 2
Canonical equation, 75
Central area tie-line power flow, 99
Computational performance, 39
Computational techniques, 1, 171
Computational time, 52
Conjugate gradient, 78
Constant head, 25
Constant multipliers, 25
Constrained minimization problem, 24
Constrained optimization, 2, 9
Control devices, 84
Control of field voltages and turbine torques, 148
Control parameters, 158
Conventional control strategy, 83
Conventional LFC approach, 69
Convergence, 40
Corner point, 103
Costate equations, 12
Cost functional, 6, 9

Damper windings, 112
Digital computers, 1
Direct axes, 185
Distributed model, 67
Dynamical constraints, 73
Dynamic optimal load flow, 2, 21
Dynamic problem, 22
Dynamic programming, 1

Economic coordination of hydrothermal–nuclear systems, 2, 55
Economic dispatch, 83
Effective fission macroscopic cross section, 58
Energy-limited constraint, 22
Euler, 5
Euler equation, 8
Exact load flow equations, 24
Exclusion equations, 64, 117
External reactivity, 58
Extrema, 5

Falling mode, 104
Feasible multilevel approach, 65
Feedback state variables, 146
Field voltages, 3, 111
Fifty-seven-bus system, 52
Fixed head hydroplants, 22

Flat frequency control (FFC), 99
Flux linkage equations, 184
Fourteen-bus system, 40
Free end point problem, 7
Frequency and tie-line power deviations, 83
Frequency error, 82
Frequency transient, 83
Fuel cost, 23
Functional analysis, 2, 3, 14
Functional analytic optimization technique, 14, 55
Functional constraint, 27

Generator inertia, 84
Global control, 132, 145, 158, 170
Governor deadband, 100
Gradient descent methods, 121
Gradient vector, 78
Gradient vector components, 102

Hamiltonian, 11, 75
Head variation, 22
Hilbert space, 16, 18, 58
Hydrothermal–nuclear systems, 55
Hydrothermal optimal power flow, 2, 22, 25
Hysteresis element, 99

Ill-conditioned equations and problems, 118, 183
Inequality constraints, 1, 11, 27
Infinite bus, 3, 111
Initial estimate, 118
Initial guess controls, 31
Initial guess estimation, 37
Initial guess vector, choice of, 32
Inner product, 15
Integral constraint, 28
Integral gain, 100
Integral part, 115
Interactive governor–supplementary regulator load frequency control, 71
Interconnected power systems, 2, 69
Interconnected turbogenerators, 111
Inverse interpolation process, 105
Iterative methods, 2
Interative techniques, 117

Jacobian, 35
Jacobian of the load flow equations, 29
Jump discontinuity, 103

Kuhn–Tucker approach, 27
Kuhn–Tucker multipliers, 114
Kuhn–Tucker theorem, 1

Lagrange multipliers, 2, 9, 114
Linear combinations of the state variables, 120
Linear regulator, 2
Linear transformation, 56
Load demand, 23
Load disturbances, 106
Load flow equations, 23
Load flow program, 21
Load frequency control (LFC), 2, 69, 83
Local control, 132, 145, 158, 168
Local minimum, 11
Loss of synchronism, 111, 120
Loss of tie-line between two areas, 91

Machine parameters, 156
Mathematical model, 112
Maximum overshoot, 83
Measurable states, 3
Mechanical power, 156
Microscopic absorption cross section, 58
Minimization of fuel cost, 9
Minimum norm, 2, 3, 14
Minimum norm formulation of functional analysis, 22, 55, 111, 120
Minimum norm problem, 63
Minimum norm theorem, 18
Mismatch vector, 34
Modified cost functional, 113
Modified Euler equation, 9
Modified mapping, 33
Monotonically increasing functional, 113
Multiarea interconnected power systems, 3, 70
Multiple corner points, 103

Natural governing characteristic, 90
Network security constraints, 22
New set point, 82
Newton iterative method, 2, 21, 22, 33
Newton–Raphson method, 105
Nodal admittance matrix, 41
Nonasymptotically stable system, 79
Noninteractive governor–supplementary regulator load frequency control, 71
Nonlinearities, 84
Nonlinear programming, 1
Nonlinear transformation, 112, 183

Norm, 15
Normed linear vector space, 15
Nuclear–hydrothermal systems, 55

Objective function, 6, 18
On power refueling capabilities, 55
Operating costs, 56
Operator, 17
Optimal control, 1, 5
Optimal control of a turboalternator, 3, 111
Optimal control parameters, 73, 83
Optimal feedback parameters, 121
Optimality conditions, 28
Optimal load or power flow, 21
Optimal parameters, 180
Optimal solution, 165
Optimal supplementary regulator, 72
Optimal tie-line control, 2, 69
Optimal torque and voltage control, 3, 111
Optimum frequency bias setting, 91
Optimum load frequency control, 3
  with governor backlash, 99
Orthogonal, 16

Parameter optimization, 142, 171
Parameter optimization problems, 3, 70
Park's transformation, 185
Penalty factor, 78, 112
Performance index weighting coefficients, 78
Piecewise-constant vector function, 77
Pontryagin's maximum principle, 1, 2, 10
Pontryagin's minimum principle, 70, 75
Prime mover, 84
Proportional and/or integral form, 70
Proportional gain, 100
Proporational-plus-integral form, 2, 69, 70, 86
Pseudo-control variables, 58, 59
Pseudoinverse operator, 19
Pseudovariable, 112, 113, 114

Quadratic form, 114
Quadratic norm, 58
Quadrature, 185

Rate of convergence, 180
Reactivity, 58
Reactor refueling cycles, 55
Realistic feedback control, 111
Realistic model, 148
Real valued positive definite scalar, 15
Regulated voltage-controlled buses, 41
Reservoir dynamics, 57
Reversing point, 104
Riccati equation, 118
Rise time, 83
Rising mode, 104
Runge–Kutta algorithms, 145, 178

Security-related constraints, 27
Series impedance, 184
Settling time, 83
Simpson's method, 183
Single area steam power system, 3
Soft constraints, 27
Sparsity, 22
Standard five-bus system, 39
State trajectory, 9, 77
Static and dynamic cases, 2
Static nonlinear equations, 21
Static optimization, 22
Stationary values, 5
Steady-state values, 112
Steepest descent method, 118
Step load change, 70
Strength of the disturbance, 158
Suboptimal strategy, 106
Successive approximation solution, 2, 27
Sudden loss of generation at one area, 94
Supplementary regulators, 71, 89
Synchronous speed, 156
Synchronous turboalternators, 3, 111
System costate equation, 76
System parameters, 78
System transient and steady-state specifications, 70

Thermal nuclear reactors, 55
Thirty-bus system, 44
Thirty-seven-bus system, 51
Three-phase short circuit, 111
Tie-line bias concept, 69
Tie-line bias controller, 91
Tie-line power disturbance, 83
Time constraints, 1
Time-coupling effect, 26
Torque and flux equations, 123
Torque angle, 112
Torque equation, 184
Torque pulse, 120
Trajectory, 10

Transformation, 17
Transient changes, 57
Transient response in the transmission line, 112
Transition matrix approach, 118
Transmission losses, 56
Transversality conditions, 5, 8
Turbine torque, 3, 111
Two fixed end points problem, 7
Two identical interconnected turbogenerators, 119
Two-level iterative approach, 65
Two mixed area interconnected power system (TMAIPS), 70
Two-point boundary value problem, 88, 117
Two separated machines, 127

Unconstrained minimization problem, 24
Unconstrained optimizaton, 2, 5
Unit commitment schedule, 25

Variable efficiency, 57
Variable head, 57
Violated limit, 29
Voltage-controlled buses, 23, 29

Water conversion factor or water-worth, 25
Water discharge, 25
Water transport delay, 57
Water volume constraints, 25

Xenon-135 concentration, 57